Physics

D. Martin Holloway.

Ordinary National Certificate Series

Editor: B. F. Gray, Hatfield Polytechnic

W. Bolton
Physics

L. W. F. Elen
Mathematics

B. F. Gray
Electrical Engineering Principles

Physics

W. Bolton
Nuffield Science Teaching Project
formerly High Wycombe College of Technology and Art

Nelson

THOMAS NELSON AND SONS LTD

36 Park Street London W1
P.O. Box 336 Apapa Lagos
P.O. Box 25012 Nairobi
P.O. Box 21149 Dar es Salaam
P.O. Box 2187 Accra
77 Coffee Street San Fernando Trinidad

THOMAS NELSON (AUSTRALIA) LTD
597 Little Collins Street Melbourne 3000

THOMAS NELSON AND SONS (SOUTH AFRICA) (PROPRIETARY) LTD
51 Commissioner Street Johannesburg

THOMAS NELSON AND SONS (CANADA) LTD
81 Curlew Drive Don Mills Ontario

First published in Great Britain 1970
Copyright © W. Bolton 1970

17 7410167

Filmset by Typesetting Services Ltd, Glasgow, and
printed by Fletcher & Son Ltd, Norwich

Contents

Part 2

Preface

This book is intended to cover the physics syllabus of the O.N.C. in Engineering. Much of it will however be suitable for the Basic Physics of the O.N.C. in Science.

The book is designed for class use by students and includes experiments, summaries of sections, worked examples and problems both numerical and for discussion. The teaching approach intended is along the lines of the Nuffield schemes where experimental work and discussion by the students forms the main teaching. Thus in this book most points are introduced by means of experiments. The text gives abbreviated accounts of the discussion after the experiments. It is intended that all the students will be doing the same experiment at the same time so that all can join in the discussion. Because of the time factor, available apparatus, and laboratory conditions, some of the experiments may have to be done as demonstrations. For virtually all the experiments the only laboratory facility required is mains supply and thus much of the work may be done in many lecture rooms. Many of the problems given at the end of the chapters are for discussion; there is no simple numerical answer or yes/no answer.

The aim of the text is a course based on experimental work and discussion by the students. It is intended to replace the traditional method of a formal lecture and a separate practical period. It is all laboratory work—where observation is all important (some of the experiments only involve observation).

This book owes much to the work of the Nuffield Physics Project—without their work this book could not have appeared in this form.

W.B.

Acknowledgments

Figs. 2–4, 2–5, 2–6, 2–7, 2–9, 2–11, 2–12, 2–13, 5–1, 11–2, 11–10, and 15–17 are taken from *PSSC Physics,* 2nd edition. Copyright © 1965, Education Development Center. Published by D. C. Heath & Co., a Division of Raytheon Education Co., Boston, Mass., U.S.A.

Figs. 1–4, 6–1, 11–4, 14–1, and 16–7 were taken from photographs supplied by Griffin & George Ltd, Ealing Road, Alperton, Wembley, Middlesex.

Figs. 2–3 and 6–6 were taken from photographs supplied by M.L.I. Ltd, 96–98 Putney High Street, London, S.W. 15.

Part 1

1 Force and Motion

1–1 Motion in a straight line

The **speed** of a body is defined as the ratio of the distance travelled to the time taken:

$$\text{speed} = \frac{\text{distance travelled}}{\text{time taken}}$$

Units: distance   metres (m)
       time   seconds (s)
       speed   metres per second (m/s)

This ratio gives the average speed for the time considered. We can only speak of the speed at some instant if we consider so small an interval of time that the speed does not change effectively over that interval.

(Experiment 1–1.)

When the speed changes, an **acceleration** is said to occur. Acceleration is the rate of change of speed, for motion in a straight line:

$$\text{acceleration} = \frac{\text{change in speed}}{\text{time taken for the change}}$$

Units: speed   metres per second (m/s)
       time   seconds (s)
       acceleration metres per second per second
                     (m/s^2)

The instantaneous acceleration can be considered as the change in speed for a very small interval of time.

If the acceleration is uniform, that is the speed is changing at a constant rate (Fig. 1–1), then a change of speed of $v-u$ occurs in time t. Thus

$$\text{acceleration} = a = \frac{v-u}{t}$$

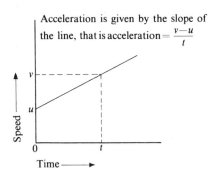

Acceleration is given by the slope of the line, that is acceleration $= \frac{v-u}{t}$

Fig. 1–1 Speed–time graph for uniform acceleration

Experiment 1–1

A ticker-tape vibrator offers us a convenient method for recording small time intervals. The vibrator presses down through a sheet of carbon paper onto a tape, leaving a mark on the tape at regular intervals, generally every 1/100 s.

Thread a tape through the vibrator and connect the machine to the appropriate power supply. Then, holding one end of the tape, walk away from the vibrator. Try the effect of walking at different speeds.

Examine the tape. Compare the spacing of the time marks, or ticks, on the sections of the tape corresponding to different walking speeds. Now cut the tape at every tenth tick. Each length of tape represents the distance you have walked in ten ticks, that is the average speed over a ten-tick interval. Compare the lengths of successive ten-tick tapes. How does the speed vary

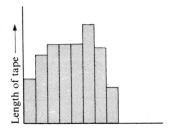

Fig. 1–2 A successive sequence of ticker tapes

where u is the speed at the beginning of the time interval, and v is the speed at the end of the time interval.

Example 1–1.   A car accelerates from rest to 20 m/s in 10 s. What is the acceleration if it is assumed to be uniform and in a straight line?

$$\text{acceleration} = \frac{v-u}{t} = \frac{20-0}{10} = 2 \text{ m/s}^2$$

During uniform acceleration the speed changes from u to v and thus the average speed during this time interval is

$$\text{average speed} = \frac{u+v}{2}$$

Example 1–2.   What is the average speed of the car in the previous example during the 10 s of motion?

$$\text{average speed} = \frac{0+20}{2} = 10 \text{ m/s}$$

Consider an object starting with a speed u and accelerating with a uniform acceleration a. At the end of one second the speed will be $u+a$ (a is the amount by which the speed changes in one second). At the end of two seconds the speed will have increased by a further

amount a and will therefore be $u+2a$. After three seconds the speed will be $u+3a$. After t seconds the speed will be $u+at$. Hence

$$v = u+at \tag{1–1}$$

As the average speed is equal to $(u+v)/2$ and is also equal to (distance travelled)/(time taken), then the distance travelled is given by

$$d = \frac{(u+v)t}{2}$$

But

$$v = u+at$$

hence

$$d = \frac{(u+u+at)t}{2}$$

$$d = ut + \tfrac{1}{2}at^2 \tag{1–2}$$

Equations (1–1) and (1–2) are called **equations of motion.** A third equation can be derived by eliminating t between (1–1) and (1–2):

$$v^2 - u^2 = 2ad \tag{1–3}$$

with time? A convenient way of showing the variation is to stick the tapes side by side, vertically, and in sequence (Fig. 1–2) on a piece of paper.

Experiment 1–2
Attach the free end of a ticker tape to a trolley which can roll down an inclined plane. Allow the trolley to roll down the plane pulling the tape through the vibrator. Cut the tape into ten-tick sections and mount successive strips side by side as in the previous experiment. How does the speed vary with time? Has the trolley accelerated uniformly? Determine the acceleration in terms of distance measured in centimetres and time measured in ten-tick intervals. Do the equations of motion apply?

Experiment 1–3
Take a photograph of a falling ball illuminated by a light

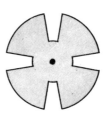

Fig. 1–3 Simple stroboscopic disk

Example 1–3. What is the distance travelled by a car moving from rest with a uniform acceleration of 2 m/s² for a time interval of 10 s?

$$d = 0 + (\tfrac{1}{2} \times 2 \times 10^2)$$
$$d = 100 \text{ m}$$

(Experiments 1–2 and 1–3.)

1–2 Force

(Experiment 1–4.) A **force** causes a change in either the shape or the motion of a body. A force is either a push or a pull, and is detected and measured by its effect on a body. These ideas enable us to talk about forces in a vague way—during our experiments we must try to discover a precise definition of force.

The most convenient conditions for us to study forces under are those in which frictional effects are very small and can be neglected. It is well known that frictional forces always oppose the motion of a body. When a force acts on a body, e.g. a push on a solid object resting on the floor, and no motion occurs it is because the applied force is balanced by an opposing frictional force: thus no motion can occur as there is no resultant force. The applied force and the frictional force tend to distort the object rather like the foam rubber in Experiment 1–4(c). We can reduce frictional forces by

floating the solid object on a cushion of gas. This can be done either by blowing air upwards through a number of holes or by using an object which supplies its own gas blowing down on a solid base. The linear air track uses the first of these methods: the carbon dioxide puck employs the second.

(Experiments 1–5 and 1–6.)

The speed of an object is the rate at which it travels a certain distance; no stipulation is made about direction. If we stipulate that the motion must be in a straight line, we use the term **velocity**. An object is said to have uniform velocity when it travels equal distances in the same straight line in equal intervals of time. A term like velocity which specifies both magnitude *and* direction is called a **vector quantity.**

Newton's laws of motion

No resultant force is necessary to keep an object at rest or moving with a uniform velocity. This rule is called **Newton's first law of motion.**

(Experiment 1–7.)

If a constant force is applied to a body, it undergoes an acceleration in the direction in which the force acts, and the magnitude of this acceleration depends on the mass of the body. If the mass is doubled, the acceleration is halved. If the mass is trebled, the acceleration is

flashing at regular intervals. The flashing light can be produced by a strobe flash unit; the same effect can be achieved by placing a rotating sectored disk in front of the camera lens (Fig. 1–3). What does the resulting picture show? Is there a uniform acceleration? What is its value?

Experiment 1–4

What is a force? What does a force do?

(a) Anchor one end of a spring and pull the other end.

(b) Push a trolley along the bench.

(c) Take a piece of foam rubber and squash it.

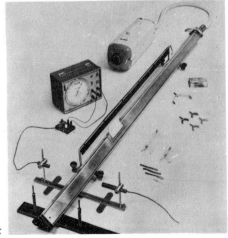

Fig. 1–4
A linear air track

reduced three times. The product of mass and acceleration is a constant.

If the force applied to an object of constant mass is increased, the acceleration increases. If the force is doubled, the acceleration doubles. If the force is trebled, the acceleration increases by the same factor.

We can represent these facts by the relation

$$F \propto ma$$

where F is the force, m is the mass, and a is the acceleration. This is **Newton's second law of motion.**

The unit of force is chosen so that

$$F = ma$$

that is

$$F = 1 \quad \text{when} \quad m = 1 \quad \text{and} \quad a = 1$$

Units: mass   kilogrammes (kg)
acceleration   metres per second per second (m/s^2)
force   newtons (N)

One point must be emphasized—the direction of acceleration is the same as that in which the resultant force acts. Force and acceleration are vector quantities and the equation is often written $\mathbf{F} = m\mathbf{a}$ to show this.

An object allowed to fall freely under the action of gravity accelerates at approximately 9.8 m/s^2. Since there is an acceleration, there must be a force of $m \times 9.8$ N acting on the object. This force is known as the **weight**; m is the **mass.** The mass of an object does not depend on its locality but the weight does, because the acceleration due to gravity varies with location on the Earth's surface.

Example 1–4. What force is necessary to produce an acceleration of 4 m/s² in a mass of 2 kg?

$$F = 2 \times 4 = 8 \text{ N}$$

Suppose that in Experiment 1–7 we had attached to the trolley two rubber bands pulling in opposite directions. If the forces were equal, no motion would have occurred. The equation $F = ma$ applies only when an unbalanced force is acting. If an object does not move under the influence of an applied force, there must be another force balancing the first. In the case of an object resting on a bench and subjected to a force acting horizontally, the opposing force is supplied by friction. (Experiment 1–8.)

Forces always occur in pairs. Whenever a force acts on one object, an equal and opposite force acts on

---

(d) Lift a weight off the floor.
Describe the forces acting in each case.

Experiment 1–5
Connect the blower to the linear air track (Fig. 1–4). Arrange the track so that it is horizontal. What happens to a carriage placed on the track and to which no force is applied? Give the carriage a gentle push. What happens? Is a force necessary to keep an object in a state of rest? Is a force necessary to keep an object moving with a uniform velocity? Photograph the movement of the carriage using a flashing strobe light or a rotating strobe disk in front of the camera.

Experiment 1–6
Place a sheet of plate glass on a level surface and wipe it clean. Fill a puck with solid carbon dioxide. This material releases a large amount of gas at room temperature which lifts the puck above the glass surface, producing almost frictionless conditions. What happens to the puck when given a push? Does it move in a straight line? Photograph the motion as in the previous experiment. Does the object move with uniform speed?

another object. In other words we can say that to every action there is an equal and opposite reaction. This is **Newton's third law of motion**.

1–3 Collisions

(Experiments 1–9 and 1–10.) In a collision momentum is always conserved. **Linear momentum** is the product of mass and velocity. Thus, in the experiments, the linear momentum of one trolley must equal the linear momentum of the other.

Example 1–5. Two trolleys fly apart when a spring is released between them. The mass of one trolley is one unit and the mass of the other is three units. The smaller trolley moves off with a velocity of 6 m/s; what is the velocity of the other trolley?

The momentum of the smaller trolley is 1×6. The momentum of the other is $3 \times v$, where v is its velocity. As momentum is conserved, we write

$$3v = 6$$
$$v = 2 \text{ m/s}$$

Momentum is conserved in any collision—the momentum before collision is always equal to the momentum after collision. This is a necessary consequence of the statement that action and reaction are equal and opposite. On collision, both trolleys must exert the same force on each other. Thus, as the forces are the same, the value of the product ma must be the same for each trolley. But acceleration is the rate of change of velocity, $a = (v-u)/t$; hence the value of $m(v-u)/t$ must be the same for each trolley. But t is the duration of the collision and this must be the same for both; hence the value of $m(v-u)$ must be the same for both:

$$-m(v-u) = M(V-U)$$
$$mv + MV = mu + MU$$

There is a minus sign on the left-hand side of the equation because the forces on the two trolleys, as well as being equal, must act in opposite directions.

Example 1–5. A mass of 0·5 kg moving with a velocity of 2 m/s strikes a mass of 2 kg which is moving along the same straight line towards it with a velocity of 1 m/s. If the 0·5 kg mass rebounds with a velocity of 2 m/s, what is the velocity of the other mass?

Take the initial direction of motion of the 0·5 kg mass to be positive. Then,

---

Experiment 1–7

If a rubber band is stretched and the extension remains constant, it is reasonable to assume that a constant force is being applied.

Loop one end of a rubber band round a pin fixed in a trolley and pull the trolley by means of the band (Fig. 1–5). One way of keeping the length of the band constant is to ensure that your hand is always above the end of the trolley. Connect a ticker tape to the trolley so that the distance the trolley moves with respect to time can be determined. To reduce friction, the trolley should be placed on a smooth slope which is inclined so that the trolley is just on the point of moving.

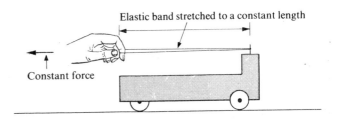

Fig. 1–5 Applying a constant force to a trolley

momentum before collision $= (0.5 \times 2) - (2 \times 1)$
momentum after collision $= (-0.5 \times 2) + (2 \times v)$

Hence

$$(0.5 \times 2) - (2 \times 1) = (-0.5 \times 2) + (2 \times v)$$

Therefore

$$v = 0$$

The 2 kg mass is stationary after collision.
(Experiments 1–11 and 1–12.)

1–4 Motion in a circle

An object will remain at rest or continue to move with a uniform velocity unless a force acts on it (Newton's first law of motion). To deflect an object from its straight-line motion a force must be applied.
(Experiment 1–13.)

For an object to move in a circular path a force must be applied at right angles to the direction of motion. This force acts towards the centre of the circle. Without this force the object will move in a straight line—if the string is released, the object moves off along a tangent to the circle.
(Experiment 1–14.)

A mass moving in a circular path can have a constant speed but *not* a constant velocity. The velocity cannot be constant as the motion is not in a straight line—a uniform velocity is a constant speed in a straight line. A force is acting on the mass and so an acceleration must be produced which acts in the direction of the force, that is towards the centre of the circle.

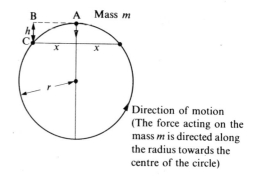

Direction of motion
(The force acting on the mass m is directed along the radius towards the centre of the circle)

Fig. 1–6 Mass moving in a circular path from A *to* C

Consider a mass m moving in a circular path of radius r with a uniform speed v (Fig. 1–6). If there were no force acting at right angles to the direction of motion, the mass would move from A to B in time t. However, because of the action of the force the mass moves from A to C during time t. The mass has moved a distance h

---

Apply a constant force to the trolley. What happens? Double the mass of the trolley by stacking two together. What effect does this have on the motion? Try three trolleys. (Check the free running of the trolleys on the slope.) If two elastic bands are extended by the same amount, twice the force is applied. What happens to the motion when two bands are used?

Experiment 1–8

Two large trolleys on which students can either sit or stand are required for this experiment. The two trolleys should be lined up facing each other with a student

seated on each. The students should hold the ends of a rope. What happens when both students pull on the rope? Where do the trolleys meet? What happens when only one student pulls on the rope?

Experiment 1–9

The trolleys required for this experiment have spring-loaded plungers. Place two trolleys on a level board with their plungers touching. Connect to each trolley a length of ticker tape so that their motion can be followed. Release the plungers so that the trolleys spring apart. Compare the velocities of the two trolleys. Repeat the experiment with the mass of one doubled by

along the direction of the force. Using the theorem of intersecting chords, we can say

$$h(2r - h) = x^2$$

or

$$2rh - h^2 = x^2$$

As h^2 is small in comparison with $2rh$, we can neglect it; thus

$$2rh = x^2$$
$$h = x^2/2r \tag{1-4}$$

If h is small, then to a reasonable approximation x is the distance travelled in time t.

$$v = x/t \qquad \text{or} \qquad x = vt$$

Substituting in (1-4)

$$h = v^2t^2/2r$$

But from the equations of motion (1-2)

$$h = \tfrac{1}{2}at^2$$

Thus

$$\tfrac{1}{2}at^2 = v^2t^2/2r$$

Hence

$$a = v^2/r$$

The force acting towards the centre is ma, or

$$F = ma = mv^2/r$$

Example 1-6. Calculate the force necessary to move a mass of 0·5 kg with a uniform speed of 3 m/s in a circle of radius 0·6 m.

$$F = \frac{0\cdot5 \times 3^2}{0\cdot6}$$

$$= 7\cdot5 \text{ N}$$

1-5 Work and energy

When a force acts on a body and displaces it, **work** is done and the force gives **energy** to the body. The work done is the product of the force (F) and the distance (d) moved in the direction in which the force is acting:

$$\text{work} = F \times d$$

Units: force   newtons (N)
        distance   metres (m)
        work   joules (J)

stacking another on top of it. Try stacking more trolleys. What can you say about the motion of the trolleys?

What quantity remains constant in the above experiment?

Experiment 1-10
An experiment similar to the previous one can be done with a linear air track. The two carriages should have taut elastic bands fixed across their bumpers. The motion of the trolleys after release can be followed with a stroboscopic photograph.

Experiment 1-11
With the aid of a linear air track study the collision of two carriages of different masses. Is momentum conserved? Is this true regardless of whether the two carriages stick together after collision or not? The carriages can be made to stick together if a pin is attached to the bumper of one and a piece of Plasticine to the other.

Experiment 1-12
Inflate a balloon and then release it so that the air can escape and the balloon can move freely. What happens? Why?

Power is the rate of doing work:

$$\text{power} = \frac{\text{work done}}{\text{time taken}}$$

Units: work   joules (J)
       time   seconds (s)
       power   watts (W)

Example 1–6.   What power is generated by a waterfall in which 10^4 kg of water fall through a height of 40 m every second?

work done = force × distance

The force is the weight of the water and this is the product of its mass and its acceleration due to gravity. Hence the work done every second, that is the power, is

$$10^4 \times 9 \cdot 8 \times 40 = 3 \cdot 9 \times 10^6 \text{ W}$$

($9 \cdot 8$ m/s^2 is the acceleration due to gravity; see Experiment 1–3).

A mass at a given height above some fixed datum position is capable of doing work in falling down to that datum position. Because it is capable of doing work, it is said to have energy. Such energy, associated with a body by virtue of its position, is called **potential energy.**

The potential energy of a body is equal to the work which the body can do when it changes its position and, in the case of a falling body, is equal to *mgh*, where *h* is the height of the mass *m* above the datum line, and *g* is the acceleration due to gravity.

When a force acts on a body and displaces it, the body is given an acceleration. There is therefore a change in velocity. If *u* is the initial velocity of the body before the application of a force, and *v* is the velocity of the body when it has travelled a distance *d* under the action of the force, then

$$\text{average velocity} = \frac{u+v}{2}$$

If *t* is the time taken to cover the distance *d*,

$$d = \frac{(u+v)t}{2}$$

Thus, the work done ($F \times d$) is

$$\frac{F(u+v)t}{2}$$

But

$$F = ma$$

and

$$a = \frac{v-u}{t}$$

In a common type of toy rocket the case is partially filled with water and the remaining air is compressed by means of a pump. If such a rocket is available, observe its motion when released. Why does it move?

Experiment 1–13

Whirl an object on the end of a length of string in a horizontal circle round your head. Is a force needed to keep it moving in a circular path? In what direction must the force be applied to the object? What happens if you let go of the string—in which direction does the object move?

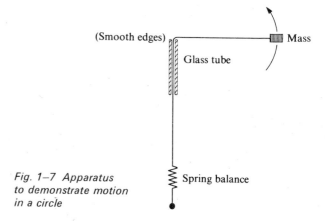

Fig. 1–7 Apparatus to demonstrate motion in a circle

(Smooth edges) Glass tube Mass Spring balance

hence

$$\text{work done} = \frac{m(v-u)}{t} \times \frac{(u+v)t}{2}$$
$$= \tfrac{1}{2}mv^2 - \tfrac{1}{2}mu^2$$

Thus the action of the force has resulted in an increase in the quantity [$\tfrac{1}{2} \times$ mass $\times$ (velocity)^2]. The work done by the force has been transformed into energy of motion, **kinetic energy**:

$$\text{kinetic energy} = \tfrac{1}{2}mv^2$$

Example 1–7. How much energy is necessary to accelerate a 1 kg mass to a velocity of 4 m/s if it starts from rest?

$$\text{energy necessary} = \tfrac{1}{2} \times 1 \times 4^2$$
$$= 8 \text{ J}$$

Summary

$$\textbf{speed} = \frac{\text{distance travelled}}{\text{time taken}}$$

The term **velocity** is used to indicate speed in a straight line.

Acceleration is the rate of change of speed in a fixed direction.

For uniform acceleration

$$a = \frac{v-u}{t}$$
$$\text{average speed} = \frac{u+v}{2}$$
$$\text{distance travelled } d = ut + \tfrac{1}{2}at^2$$

No force is necessary to keep an object at rest or moving with a uniform velocity (constant speed in a straight line) (**Newton's first law of motion**).

An unbalanced force acting on a mass produces an acceleration (**Newton's second law of motion**).

$$F = ma$$

Balanced forces can produce a change in shape of the mass but no acceleration. When a force acts on a body there must be an equal and opposite force acting somewhere else: to every action there is an equal and opposite reaction (**Newton's third law of motion**).

Linear momentum is the product of mass and velocity, and is always conserved.

Motion in a circle is obtained by the action of a force at right angles to the direction of motion:

$$F = mv^2/r$$

Work is done when a force displaces its point of application and is equal to the product of the force and

---

Experiment 1–14

Attach a piece of string to a small mass and pass the other end of the string through a glass tube (Fig. 1–7), the ends of which must be very smooth. Tie the free end of the string to a spring balance fixed to the floor or some other immovable object. Whirl the mass on the end of the string in a horizontal circle round your head, holding the glass tube vertical. Keep the force constant and, by measuring the time taken for a number of revolutions, determine how the speed of the mass is related to the radius of the circle.

Now vary the force for a fixed radius, and determine how the three variables—force, velocity, and radius—

are related. If time permits, try different masses on the end of the string.

the distance of displacement in the direction of the line of action of the force:

work $= F \times d$

Power is the rate of doing work, that is work done divided by time taken.

Potential energy is the energy associated with a body by virtue of its position (with respect to a fixed datum position):

potential energy $= mgh$

Kinetic energy is the energy associated with a body by virtue of its motion:

kinetic energy $= \frac{1}{2}mv^2$

Units: distance metres (m)
        mass   kilogrammes (kg)
        time   seconds (s)
        force   newtons (N)
        energy   joules (J)
        power   watts (W)

Problems

1–1 A car travels at 40 km/h; how far does it travel in 15 min?

1–2 Draw a graph showing how the distance of an object from a fixed point varies with time when it is (a) moving with a uniform speed, and (b) uniformly accelerating.

1–3 A car accelerates uniformly from 10 m/s to 20 m/s in 10 s. What is its acceleration?

1–4 An object falls freely from rest with an acceleration of 9·8 m/s². What distance will it cover in the first 10 s of fall?

1–5 What is the distance travelled by a car moving from rest with a uniform acceleration of 3 m/s² for a time interval of 10 s?

*1–6 What is the effect of a force acting on a body?

1–7 What force is necessary to give an object of mass 2 kg an acceleration of 5 m/s²?

*1–8 Explain how a force may or may not be necessary to keep an object moving along a horizontal surface.

*1–9 Would you expect Newton's laws of motion to be different on another planet?

1–10 A particle of mass 100 g collides with a stationary particle of mass 200 g. If the particle has a velocity of 10 cm/s before the collision and 3·3 cm/s in the same direction afterwards, calculate the velocity of the 200-g particle after the collision.

1–11 A particle of mass 10 g moving with a velocity of 30 m/s normal to a wall, strikes it, and rebounds with the same velocity. What is the force experienced by the wall if the collision lasts 1/50 s?

1–12 What forces would you experience if you were in a train (a) moving with constant velocity in a straight line, (b) accelerating in a straight line, and (c) moving with a constant speed round a curve?

*1–13 After a rocket has been launched it is decided to explode it.
(a) What is known about the momentum of the fragments?
(b) What is known about the momentum of the centre of mass of the rocket?

*1–14 At fairgrounds there is often a large machine called a rotor. This consists of a large cylindrical drum with its axis vertical. People stand in the cylinder and when it is rotated they 'stick' to the walls. How can this be explained?

1–15 A mass of 0·5 kg is swung round in a horizontal circle on the end of a cord of length 2 m with a constant speed. If the speed of the mass is 10 m/s, what tension must exist in the cord? What would happen if the cord broke?

1–16 A particle of mass 10 g moving with a velocity of 10 m/s is completely stopped on encountering a wall. How much energy is lost by the particle?

1–17 How much work is necessary to lift a mass of 2 kg through a vertical distance of 3 m?

2 Wave Motion

2–1 Waves

If a stone is dropped into a pool of water, a disturbance moves out across the water surface; if one end of a rope is moved up and down vigorously, a disturbance moves along the rope. In both cases the disturbance indicates a movement of energy by means of a wave. The particles of which the rope is made move up and down; they do not move along the rope. There is only a movement of energy. A cork floating on the water bobs up and down as the wave passes; it will not move across the surface.

In both of these examples of wave motion the particle displacement is at right angles to the direction in which the wave is moving. These are called **transverse waves.** In the case of the rope there is more than one possible plane of vibration in which the wave can occur (Fig. 2–1).

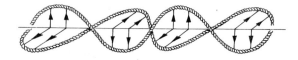

Fig. 2–1 Possible transverse vibrations of a rope

The speed at which a wave crest or trough moves along the rope or the water surface is known as the **wave velocity** (v). The distance between two successive crests or troughs is known as the **wavelength** (λ). The number of wavelengths produced per second by the disturbance is known as the **frequency** (f) (Fig. 2–2). These three quantities are related.

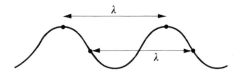

Fig. 2–2 The term wavelength

In one second, f waves are produced and, if the wavelength is λ, then in that time the wave will have reached a point $f \times \lambda$ distant from the source. The velocity is the distance travelled divided by the time taken; thus

$$v = f\lambda$$

Experiment 2–1

Fix one end of a long coil spring to a rigid support and pull the other end just enough to keep the spring taut. Send a transverse pulse along the spring by moving one end quickly from side to side. Observe the pulse both before and after its encounter with the fixed end.

Is the pulse reflected? Is there any change in the pulse after reflection?

Experiment 2–2

Instead of using a rigid support to fix one end of the spring, attach the end to a length of cotton, and fix the

other end of the cotton. Send a transverse pulse along the spring as before.

What happens when the pulse reaches the junction of the spring and cotton? Does reflection occur? Are the results the same as for reflection at a fixed boundary?

Experiment 2–3

Water waves are most conveniently studied in a ripple tank (Fig. 2–3) which is a shallow rectangular water trough mounted on legs. The base of the trough is transparent and, when a car headlamp bulb is placed about 70 cm above the water surface, images of the waves are produced on the floor beneath the tank (the

Units: v metres per second (m/s)
$\quad\quad f$ hertz (Hz)
$\quad\quad\quad$ cycles per second (c/s)
$\quad\quad \lambda$ metres (m)

This is a general equation which applies to all forms of wave.

Example 2–1. A water wave has a wavelength of 0·5 cm and is produced by a source vibrating at 50 Hz. Calculate the wave velocity.

$$v = f\lambda$$

Hence

$$v = 50 \times \frac{0\cdot5}{100}$$
$$\quad = 0\cdot25 \text{ m/s}$$

2–2 Reflection of waves

The fact that particles and certain objects such as table tennis balls can be 'reflected' is obvious, but can waves be reflected? Unless a heavy rope is used, it is not easy to study the reflection of waves along a rope. Rubber tubing filled with sand or a long coil spring ('slinky' spring) are generally more convenient.

(Experiments 2–1 and 2–2.) The experiments show

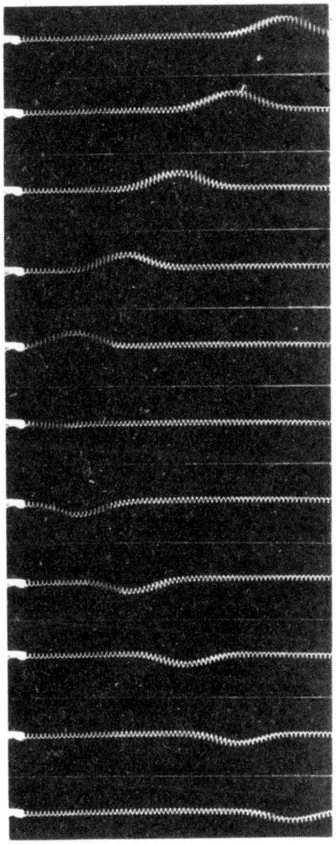

Fig. 2–4 Reflection
of a pulse from a fixed end

Fig. 2·3 A ripple tank

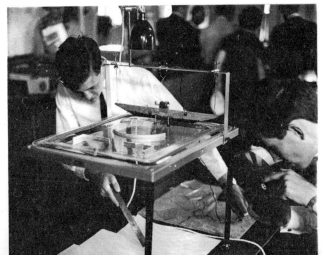

waves act as lenses). Waves can be produced by an electrically driven vibrator or, if pulses only are required, by allowing a drop of water to fall into the trough (circular waves) or by rolling a cylindrical rod in the water with a short sharp motion (plane waves).

Place a strip of metal in the tank and study the reflection of both circular- and plane-wave pulses from it. Is there any relationship between the angles of incidence and the angles of reflection? Do the waves move in straight lines? (The angle between the direction of the incident wave and the normal to the surface is called the **angle of incidence** and the angle between the reflected wave and the normal is the **angle of reflection**.

that reflection of waves occurs at both fixed and free boundaries. There is, however, a difference: when the boundary is fixed, the reflected pulse is upside down (Fig. 2–4); when the boundary is free, the reflected pulse is the same way up as the incident pulse (Fig. 2–5). We say that reflection at the fixed boundary has produced a **phase change,** that is the waves are now 'out of step'. No phase change has occurred at the free boundary. The phase change at the fixed boundary is equivalent to a displacement of wavelength of $\frac{1}{2}\lambda$, that is the wave has advanced by half a wavelength on reflection.

The experiments with the spring involve incident and reflected waves that can travel along one direction only; when waves are produced on water, however, the incident and reflected waves need not travel along the same direction.

(Experiment 2–3.) When water waves are reflected no change of wavelength, and hence velocity, occurs; this can be easily observed in a ripple tank (Fig. 2–6). In addition there is a simple relationship between the directions of the incident and reflected waves—the angle of incidence is equal to the angle of reflection, the angles being measured between the direction of wave propagation and the normal to the reflecting surface.

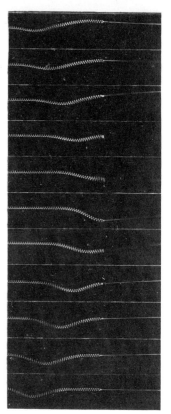

Fig. 2–5 A pulse on a spring reflected from a junction with a very light thread

Example 2–2.   A water wave is incident on a barrier at an angle of 30°. Calculate the angle of reflection.

angle of incidence = angle of reflection

Therefore, the angle of reflection must be 30°.

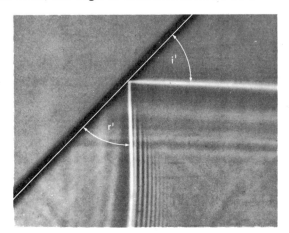

Fig. 2–6 The reflection of water waves

2–3 Refraction

The wavelength of a wave in the ripple tank can be measured directly if the water surface is illuminated with a light flashing at the same frequency as that of the water waves. The light from a car headlamp bulb can be made to pass through a rotating disk in which holes have been cut or a stroboscope can be used. The effect

Fig. 2–7 Refraction of waves at a boundary between deep and shallow sections of the ripple tank. Note the weak reflected waves

of both these methods is to give an apparently stationary view of the waves because the waves are only viewed when they occupy certain positions.

(Experiment 2–4.) When a wave on the surface of shallow water travels from a region of one depth to a

---

Experiment 2–4
Measure the wavelength of the waves produced by the electrically operated plane-wave vibrator.

Lay a piece of glass in the bottom of the tank so that the water waves pass abruptly from a deep to a shallow region. What happens to the waves? Is there a change in direction? Is there any change in wavelength?

region of different depth, **refraction** occurs, that is the direction of travel of the wave motion changes. The change in direction is produced because a change in velocity occurs; this is shown by a change in wavelength (Fig. 2–7). When a wave passes from a region of higher velocity to one of lower velocity, the direction of wave motion is inclined more towards the normal. The ratio of the velocities is called the **refractive index** (μ).

$$\mu = v_1/v_2$$

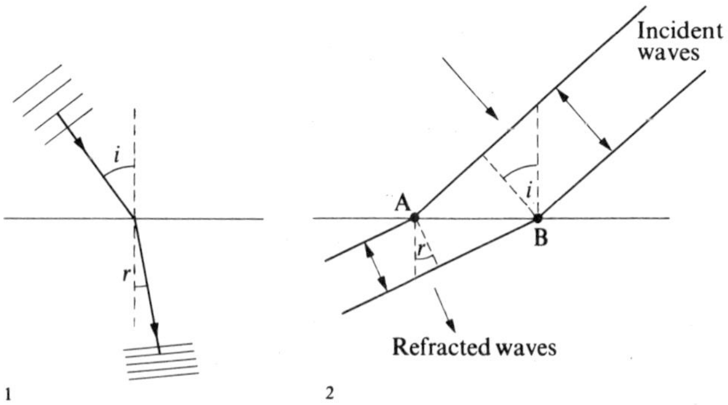

Fig. 2–8 Refraction

The refractive index is also equal to the ratio of the sine of the angle of incidence to the sine of the angle of refraction. From Fig. 2–8

$$\sin i = \lambda_1/\mathrm{AB} \qquad \text{and} \qquad \sin r = \lambda_2/\mathrm{AB}$$

Hence

$$\frac{\lambda_1}{\lambda_2} = \frac{\sin i}{\sin r}$$

But

$$v_1 = f\lambda_1 \qquad \text{and} \qquad v_2 = f\lambda_2$$

Thus

$$\text{refractive index } \mu = \frac{v_1}{v_2} = \frac{\sin i}{\sin r}$$

Example 2–3. A wave, moving with a velocity of 10 cm/s, approaches a change in depth at an angle of incidence of 30°. If the wave moves at an angle of refraction of 20° in the new level, calculate the new velocity.

$$\frac{\sin i}{\sin r} = \frac{\sin 30°}{\sin 20°} = \frac{0.50}{0.34} = 1.47$$

Hence

$$\text{new velocity} = \frac{10}{100 \times 1.47} = 0.068 \text{ m/s}$$

2–4 Interference

When two table tennis balls meet they bounce off each other; what happens when two waves meet?

Experiment 2–5

Two people should hold a large spring on the bench so that it is slightly extended. Each should direct, at the same instant, a pulse along the spring. What happens when the pulses meet? What happens after the pulses have met?

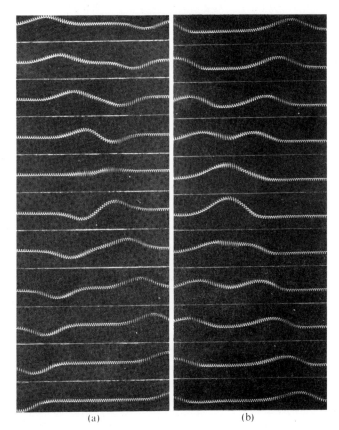

(a)   (b)

Fig. 2–9 (a) *The superposition of two equal and opposite pulses* (b) *The superposition of two equal and symmetric pulses*

(Experiments 2–5 and 2–6.) When two water or 'spring' waves meet, addition of the two displacements occurs at every point (Fig. 2–9). If two crests coincide, a bigger crest is produced—if a crest and trough coincide, then cancellation occurs. Waves in phase, that is in step with crest coinciding with crest and trough with trough, produce bigger displacements—waves out of phase, that is out of step, produce smaller or zero displacements. With two point sources producing waves of the same amplitude and frequency, maximum displacements occur at those points where waves meet in phase,

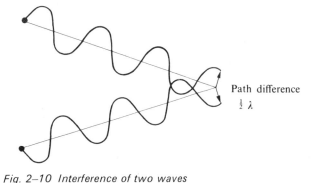

Fig. 2–10 Interference of two waves

Path difference $\frac{1}{2}\lambda$

Experiment 2–6

Remove the plane-wave vibrator from the ripple tank, and clamp two prongs about 6 cm apart and just touching the water surface. With the motor running, two point sources of waves will be produced. What happens when the waves meet? Use a strobed light to observe the result.

Change the wavelength of the waves by altering the voltage applied to the motor. What happens to the wave pattern?

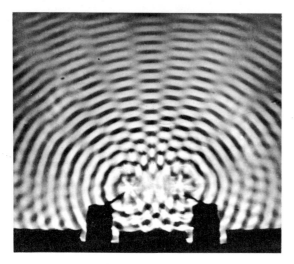

Fig. 2–11 Interference pattern from point sources in phase

point to be a maximum is that the path difference is a whole number of wavelengths. For maximum displacement

$$\text{path difference} = n\lambda$$

where $n = 0, 1, 2, 3, 4$, etc. (an integer). For zero displacement the two waves must have a path difference of an odd number of half wavelengths. For zero displacement

$$\text{path difference} = (n + \tfrac{1}{2})\lambda$$

and zero displacements occur where there is a phase difference corresponding to one-half of a wavelength (Fig. 2–10). The resulting pattern is known as an **interference pattern** and is characteristic of wave motion (Fig. 2–11).

If two waves of equal amplitude and frequency are in phase at their sources and the displacements are collinear, then the only condition for the resultant at a

2–5 Standing waves

It is possible to obtain an interference pattern between the waves from a source and their reflections from some surface. The waves generated by a vibrator are travelling waves which transfer energy from one point to another. When interference occurs between two travelling waves, a stationary pattern can be produced and the waves are called **standing waves.**

Look again at the pattern produced by the two vibrators in the ripple tank: though the waves move, the interference pattern is stationary (no stroboscope is necessary to stop the motion of the interference region). It is particularly easy to show these stationary or standing waves if the interference is between the waves

---

Experiment 2–7
Place a plane reflector parallel to the plane vibrator in a ripple tank. Explain the resulting interference pattern. How far apart are the interference fringes?

incident on a reflector and those reflected from it. For the waves incident along the normal to a plane reflector the stationary waves form a pattern of lines parallel to the reflector.

(Experiment 2–7.)

Example 2–4. In a standing-wave pattern, maxima are found 2 cm apart. What is the wavelength of the waves producing the standing waves?

For standing waves the distance between successive maxima is equal to half the wavelength ($\frac{1}{2}\lambda$). In this case

$\frac{1}{2}\lambda = 2$ cm

Thus

$\lambda = 4$ cm

Hence the wavelength is 4 cm.

2–6 Diffraction

When a stream of particles is incident on a small hole or slit in a screen, a narrow well-defined beam of particles emerges. This occurs whether the slit is wide or narrow. What happens with waves?

(Experiment 2–8.) When a narrow parallel beam of waves passes through a slit in a screen, the waves spread

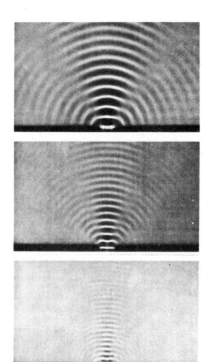

Fig. 2–12 Three views of waves passing through the same opening

Experiment 2–8
Assemble the ripple tank with the plane-wave source and barriers so placed that the waves are incident along the normal to a slit about 10 cm wide. What happens to the waves as they pass through the slit? Reduce the width of the slit. How does the wave pattern change as the slit width is reduced? What pattern is formed when the slit width is smaller than the wavelength?

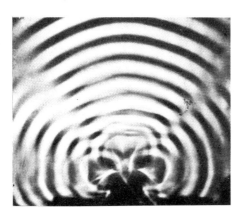

out, that is instead of emerging from the slit in a parallel beam they diverge from the slit. This effect is called **diffraction,** and is characteristic of waves. It becomes particularly noticeable when waves pass through a slit whose width is about equal to the wavelength concerned (Fig. 2–12). When the slit is one wavelength in width the waves spread out as though the slit had been replaced by a point source. When wider slits are used, the effect is that of a number of point sources (Fig. 2–13). Diffraction also occurs when waves impinge on an object placed in their path, as well as when they pass through an aperture. An obstacle less than about one wavelength wide hardly affects the waves, no 'shadow' being cast.

Example 2–5. What wavelength is necessary for an obstacle about 3 cm wide to cause diffraction?

A wavelength less than about 3 cm.

Fig. 2–13 A diffraction pattern of straight waves passing through a slit and, below, an interference pattern of a line of equally spaced point sources extending across the slit. Near the sources the effect of source separation leads to some difference in the patterns. Far away the two patterns are the same.

The diffraction pattern produced by slits wider than one wavelength shows the effects of diffraction and interference, the slits being effectively replaced by a number of point sources. Interference occurs between the waves emitted from each of these sources. The resulting pattern is the combination of these effects.

Summary

Waves can show **reflection, refraction, interference,** and **diffraction.** Interference and diffraction are properties peculiar to waves and are not shown by particles.

Frequency is the number of waves produced per second by a source; **wavelength** is the distance between successive corresponding points on a wave, e.g. successive crests. The **wave velocity** is the product of the frequency and the wavelength. Refraction of water waves is caused by a change in wave velocity which occurs when a wave crosses a boundary between two different depths of water. The **refractive index** is the ratio of the two velocities.

Interference occurs when waves meet. It can result in either addition or subtraction of the particle displacements, that is reinforcement or cancellation. Waves diffract, that is spread out and interfere, on passing through an aperture or around an object.

Problems

2–1 Consider a point vibrator dipping into water in a ripple tank.
  (a) In which direction is a cork on the water surface displaced by the ripples?
  (b) In which direction does the energy of the wave move?

2–2 A wave is incident on objects of size 0·1, 1, and 100 times the wavelength. What is the effect of these objects on the wave?

2–3 (a) What happens when waves 'collide'?
  (b) What happens when particles collide?

2–4 A wave has a wavelength of 5 mm and a frequency of 20 Hz. What is the velocity of the wave?

2–5 A wave of length 3 mm passes from one medium to another. If the angle of incidence of the wave on the new medium is 45° and the angle of refraction 30°, calculate the wavelength in the new medium.

2–6 Would you increase or decrease the frequency of a ripple generator if an increased wavelength was required?

*2–7 Assume that the image of a regular water wave has been 'frozen' by means of a stroboscope. What would you see if the stroboscope frequency was doubled?

2–8 A wave approaches (a) a straight barrier at an angle of incidence of 30°, and (b) a convex circular barrier. What happens on reflection?

2–9 Draw a scale diagram for waves of length 1 cm emitted from two points a distance 3 cm apart, and hence determine the positions where constructive and destructive interference occur.

2–10 Standing waves with maxima 2 cm apart are produced by waves incident normally on a reflector. What is the wavelength?

2–11 Waves, of length 2 mm, are incident on slits of width 2 mm, 6 mm, and 20 mm. At which slit will diffraction be most pronounced?

2–12 At what positions on a screen 40 cm from two vibrators, and parallel to a straight line joining the two, will constructive and destructive interference occur? The two vibrators are in phase and produce waves of length 5 mm, the separation of the vibrators being 2 cm.

3 Vibrations

3–1 Simple harmonic motion

(Experiment 3–1.) Consider an oscillating system such as a simple pendulum or a spring suspended at one end and carrying a weight at the other. Such systems when set in motion will oscillate with a gradually diminishing amplitude, finally coming to rest. During the oscillation there is a continuous interchange between the kinetic and potential energies of the system. In addition there is an energy loss due mainly to frictional forces and air resistance. The energy is not really lost—it is lost only as far as the vibrating system is concerned and is, in fact, converted into heat, another form of energy.

A system that loses energy in this way is said to have its motion **damped.** If there were no damping—an ideal situation which cannot be realized in nature—the motion of the systems mentioned above would be of constant amplitude, that is the vibrating mass would travel the same distance on either side of its rest position. (Experiments 3–2 and 3–3.)

Systems whose displacement–time graph is a sine wave of constant amplitude are said to oscillate with **simple harmonic motion** (Fig. 3–2). We can also specify simple harmonic motion in terms of acceleration: the acceleration is always directly proportional to the distance of the mass from its rest position and is always

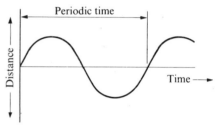

Fig. 3–2 Distance–time graph (sine curve)

Experiment 3–1

Observe a number of oscillating systems; a simple pendulum, a loaded spring, and a trolley tethered by elastic bands between two points on a horizontal surface (Fig. 3–1).

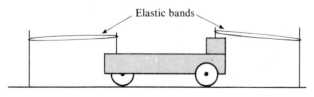

Fig. 3–1 Arrangement for oscillations of a trolley

The amplitude of the oscillations diminishes with time—what happens to the frequency? Is the frequency of oscillation the same each time the experiment is repeated? What happens to the energy of the system?

Experiment 3–2

Construct a paper funnel which can be attached to the end of a long pendulum and which will allow a steady stream of sand to flow from it as the pendulum oscillates. With the pendulum oscillating steadily, pull a sheet of paper along underneath it. The resulting pattern shows how the position of the end of the pendulum varies with time. Does the trace produced look typical of vibrations?

Experiment 3–3

Take a stroboscopic photograph of the motion of a pendulum bob. Is the velocity of the bob constant? Sketch a graph showing how the velocity depends on the position of the bob.

Acceleration is the rate of change of velocity (see Chapter 1)—is there an acceleration in this case and, if so, what is its direction? From your velocity–distance graph obtain a graph of acceleration against distance.

The experiment could be performed with any of the

directed towards the rest position, that is the mass accelerates towards its rest position and decelerates or experiences a negative acceleration as it moves away from it. Simple harmonic motion is a very common form of motion. Thus the displacement $y = A \sin \omega t$, where A is the amplitude, that is the maximum displacement, which is constant, and ω is a constant. The velocity of the mass is its rate of change of position, that is dy/dt. Hence the velocity is given by

$$\frac{dy}{dt} = A\omega \cos \omega t$$

Verify this from the displacement–time graph and the velocity–time graph obtained in Experiment 3–2. The acceleration is the rate of change of velocity, that is dv/dt. Hence acceleration is given by

$$\frac{dv}{dt} = \frac{d^2y}{dt^2} = -A\omega^2 \sin \omega t$$

But $y = A \sin \omega t$ and hence acceleration is given by

$$\frac{d^2y}{dt^2} = -\omega^2 y$$

This again can be checked by examining the results of Experiment 3–2.

The time taken for one complete oscillation is called the **period** T. This is also the time for one complete cycle of the displacement–time graph (Fig. 3–2), and corresponds to a change in ωt of 360° or 2π radians. Hence the period is given by

$$\omega T = 2\pi \qquad \text{or} \qquad T = 2\pi/\omega$$

The **frequency** f of the motion is the number of complete periods in one second:

$$f = 1/T = \omega/2\pi$$

Now, since ω^2 is the acceleration for unit displacement from the rest position, we have

$$T = \frac{2\pi}{\omega} = \frac{2\pi}{(\text{acceleration for unit displacement})^{1/2}}$$

other oscillating systems—a particularly convenient one is the trolley tethered between two vertical supports by elastic bands. A vertical straw attached to the trolley enables the measurements to be easily made on the photograph.

Experiment 3–4
Measure the periodic time of a tethered trolley. The constant K in the expression for periodic time can be obtained by suspending one of the elastic bands vertically and by measuring its extension when various weights are attached to its lower end. A graph of weight against extension will have a slope equal to K. Hence check the validity of the expression.

Experiment 3–5
Determine the factors affecting the frequency of oscillation of a simple pendulum, e.g. the length of the string, the mass of the bob, and the amplitude of the oscillation.

Experiment 3–6
Determine the factors that affect the frequency of oscillation of a steel rule or metal strip clamped at one end.

Experiment 3–7
Charge a capacitor by applying a d.c. voltage across it. Discharge the capacitor through an inductance, and a meter connected in series (Fig. 3–3). What happens? How can the frequency be altered?

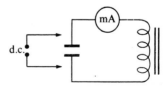

Fig. 3–3 Discharge of a capacitor through an inductor

Therefore

$$T = 2\pi\left(\frac{\text{displacement}}{\text{acceleration}}\right)^{1/2}$$

Consider the tethered trolley experiment (Experiment 3–1). The extension of the rubber bands is proportional to the applied force:

force $= K \times$ extension

where K is some constant.

When the trolley is displaced to one side of the central rest position the extension of one band is increased from X to $X+x$ and the other band has its extension decreased from X to $X-x$. Thus the two forces acting on the trolley are

$K(X+x)$ and $K(X-x)$

The resultant force is therefore

$K(X+x) - K(X-x) = 2Kx$

This is the force causing the acceleration and thus

$2Kx = \text{mass} \times \text{acceleration}$

Hence

periodic time $T = 2\pi\left(\dfrac{\text{mass}}{2K}\right)^{1/2}$

In the mechanical systems there is a continual interchange of potential and kinetic energy. In the case of the capacitor and inductor, energy is alternately stored in the electric and magnetic fields respectively. In all these practical cases energy is dissipated and the oscillations die away.
(Experiments 3–4, 3–5, 3–6 and 3–7.)

3–2 Natural and forced vibrations
(Experiments 3–8 and 3–9.) When the applied frequency is the same as the natural frequency of an oscillating system, oscillations of large amplitude are produced in

---

The capacitance of the capacitor should be in the microfarad range and the inductor should have about 20,000 turns or more of wire on a C core. A 2-0-2 mA meter can be used.

Experiment 3–8
Place the tethered trolley on a board which can be rocked on a roller along its centre line (Fig. 3–4). Rock the board at various frequencies both above and below the natural frequency of the trolley–elastic-band system. What happens? Measure the amplitude of the

trolley oscillations at the different applied frequencies and plot a graph of amplitude against frequency.

Experiment 3–9
Tie a taut length of string horizontally between two supports (Fig. 3–5); suspend a mass of 1 kg from it on a string 1 m long and a mass of 100 g on a 70-cm string

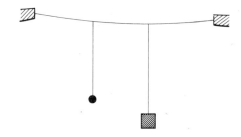

Fig. 3–5 Apparatus for the study of the effect of an applied frequency on the oscillations of a pendulum

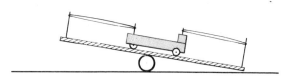

Fig. 3–4 Forced oscillations of a trolley

the system; this is called **resonance.** At all other frequencies the amplitude is less (Fig. 3–6). All structures have a resonant frequency and, when the applied frequency is equal to the resonant frequency, a large amplitude results. For example, an aircraft has a number of resonant frequencies and care must be taken to see that none of the frequencies are realized during normal operations, otherwise the build-up of oscillations could lead to serious damage. A suspension bridge across the

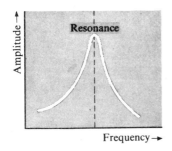

Fig. 3–6 Variation of amplitude with frequency

Tacoma Narrows in the U.S.A. collapsed in 1940 when wind gusts set it swinging at its resonant frequency.
(Experiments 3–10 and 3–11.)

3–3 Standing waves

(Experiment 3–12.) **Standing waves** are produced by interference between incident and reflected waves. In the case of the string, different wave patterns are produced depending on the frequency used. When the string forms a single loop, the frequency is known as the **fundamental** (Fig. 3–8). The frequency at which two loops are formed is known as the **first overtone** or **second harmonic frequency.** At yet higher frequencies more harmonics are produced.
(Experiment 3–13.)

In all standing-wave patterns a **node** (point of no displacement) is produced at a fixed boundary and an **antinode** (point of maximum displacement) at a free end. Thus in the case of the tube open at one end the possible standing-wave patterns that can exist must have an

from a point a short distance away along the horizontal string. The natural frequency of a pendulum depends only on its length; thus, when the heavy pendulum is set in motion, it does so with a frequency related to its length. This frequency will be imposed on the motion of the light pendulum. In what way does the maximum amplitude of the light pendulum depend on the length of the heavy pendulum? This is the same as considering how amplitude depends on applied frequency.

Experiment 3–10
Suspend a length of rope between two supports so that it hangs loosely between them. Now blow on the rope. Try blowing in short puffs. Can you set up resonance?

Experiment 3–11
Connect a capacitor and inductor in series with a signal generator (Fig. 3–7). Measure the current in the circuit

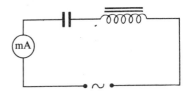

Fig. 3–7 Apparatus for the study of the effect of an applied frequency on the oscillations of a capacitor–inductor combination

with an a.c. meter for various values of frequency. Plot a graph of current against applied frequency. How could you determine the natural frequency of the capacitor–inductor system?

Experiment 3–12
Attach one end of a light horizontal string to a vibration generator and pass the other end of the string over a

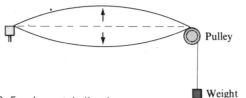

Fig. 3–8 Fundamental vibration

antinode at the open end and a node at the closed end (Fig. 3–9). Standing waves occur when the applied frequency coincides with the natural frequency, or one of them, of the system concerned. In the case of systems

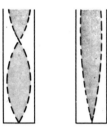

Fig. 3–9 Standing waves in a column of gas
(Note. The vibrations are in fact longitudinal and the diagrams represent a displacement–distance graph)

such as strings and tubes, the resonant frequency depends on the length of the vibrating object, that is the length of the string or the column of air in the tube. (Experiments 3–14 and 3–15.)

Summary
All structures have natural frequencies of vibration. Many systems vibrate with almost **simple harmonic motion.** When this occurs the displacement of a point of the system varies with time according to a sine curve. One way of describing simple harmonic motion is to state that a system, whose acceleration is proportional to the distance from the rest position and always directed towards it, moves with simple harmonic motion. The **periodic time,** the time for one complete oscillation, is given by

$$T = \frac{2\pi}{\omega} = 2\pi \left(\frac{\text{displacement}}{\text{acceleration}}\right)^{1/2}$$

Forced oscillations occur when a system is caused to

pulley to a mass of about 250 g (Fig. 3–10). Observe what happens as the frequency applied to the generator is varied.

The generator could be a simple one based on a length of metal being magnetized alternately in different directions when between the poles of a horseshoe magnet (Fig. 3–11).

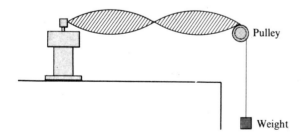

3–10 Apparatus for the study of standing waves on a string

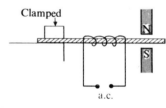

Fig. 3–11 Simple vibrator

Experiment 3–13
Place a loudspeaker which is connected to an audio-frequency signal generator over the open end of a tube, the other end being closed. What happens if the frequency of the signal generator connected to the loudspeaker is altered? What is the fundamental frequency of the tube?

oscillate at a frequency other than its own natural or resonant frequency. If the applied frequency equals the natural frequency of the system, oscillations of large amplitude occur. When a system is allowed to oscillate freely, it does so at its natural frequency.

Problems

3–1 Explain what happens when the frequency applied to an object is equal to its natural frequency.

3–2 What is simple harmonic motion?

3–3 Derive an equation for the natural frequency of an oscillating spring.

3–4 In simple harmonic motion when do the following occur: (a) maximum acceleration, (b) no acceleration, (c) maximum velocity, and (d) zero velocity.

3–5 What is the resonant frequency of a tube 50 cm long which is open at one end and closed at the other? The speed of sound in air is 340 m/s.

3–6 What effect will there be on the frequency of the tube in the previous question if both ends are open?

Experiment 3–14

Clamp a thin metal plate at its centre and sprinkle a thin layer of lycopodium powder or fine sawdust over it. Set the plate into oscillation by gently tapping one corner. The powder will move to regions of no displacement (nodes) and hence will show the wave pattern produced in the plate.

Experiment 3–15

Study the behaviour of a vibrating circular disk clamped around its periphery. The disk behaves in a similar manner to the diaphragm of a telephone earpiece. Apply different frequencies to the centre of the disk by means of a vibrator. Can you find a resonant frequency? Are these frequencies normally encountered in speech? An alternative is to apply different frequencies to the earpiece of a telephone or headphones with an audio-frequency signal generator and determine by observation of the earpiece disk when resonance is reached; the volume of the note will increase considerably at resonant frequencies.

4 Sound

4–1 Longitudinal waves

(See Experiment 4–1.) When the coils of a spring or the particles of a substance are squeezed together we talk of a **compression**, while, if the coils or particles are separated, it is termed a **rarefaction**. A pulse or a wave travelling along the spring does so by means of compressions and rarefactions (Fig. 4–1); such a wave is known as a **longitudinal wave**, the particle displacement being along the line of wave motion. In common with transverse waves we can refer to the frequency and wavelength of the wave.

Example 4–1. The wavelength of a disturbance travelling along a spring was measured as 30 cm when the frequency was 5 Hz. What was the speed of the wave along the spring?

$$v = f\lambda$$

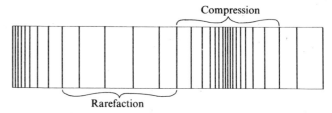

Fig. 4–1 A longitudinal wave

Hence

$$v = 5 \times 30 = 1 \cdot 5 \text{ m/s}$$

(Experiment 4–2.)
As with transverse waves, interference can occur between two longitudinal waves.

4–2 Sound waves

(See Experiment 4–3.) Whenever sound is heard, something is vibrating, the cone of the loudspeaker, the

---

Experiment 4–1
Extend a long 'slinky' spring and hold it on a flat horizontal surface. Give a sharp push to the spring coils at a point near one end so that they are temporarily compressed there. Explain what happens.

What is the direction of motion of the pulse? What is the direction of motion of the coils of the spring?

Move the end of the spring backwards and forwards in a regular manner. What happens?

Experiment 4–2
Extend the spring again. What happens if longitudinal pulses are sent along the spring from both ends simultaneously? What happens if waves are sent from both ends?

How do results compare with those produced with transverse waves? Transverse waves can be produced by giving the end of the spring an up-and-down motion.

Experiment 4–3
How is sound produced? Examine as many different means as you can of producing sound.

Gently touch the cone of a loudspeaker while sound is being produced. Put the loudspeaker horizontal and place a few beads or other light objects on the cone. What happens to them? Is there any difference if the volume is increased?

How is sound produced from a guitar or violin? How is sound produced from a drum? How does a tuning fork produce sound?

Experiment 4–4
Connect a loudspeaker to a signal generator and vary the frequency. Can you hear all the frequencies? What are your frequency limitations? Has everybody in the class the same limitations?

string of the guitar, the skin of the drum, or the prongs of the tuning fork. The oscillations are similar to those producing longitudinal waves in the spring, a backward and forward motion. This motion causes the air molecules to be pushed and pulled in much the same way that the coils of the spring were pushed and pulled. Each air molecule oscillates about a mean position in the same way that the coils of the spring each oscillated about a mean position. **A sound wave** is a longitudinal wave with the sound energy being propagated by means of compressions and rarefactions. Not all such waves are audible to the human ear.

(Experiment 4–4.) The **sensitivity** of the ear depends on the individual and is generally greatest for a young person, the frequency limits being about 20 Hz to 20,000 Hz. Sounds with frequencies beyond the upper limit of audibility are called **ultrasonic.**

(Experiment 4–5.) A musical sound consists of a repetitive wave pattern (Fig. 4–2). It has a definite wavelength and a definite frequency—but what is heard

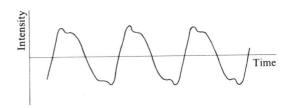

Fig. 4–2 Wave-form obtained when a piano key is struck

depends on the listener. If you were tone deaf, all frequencies would sound the same. The characteristic of a note as determined by a listener is known as the **pitch** of the note. Generally the higher the frequency is, the higher the pitch.

4–3 The speed of sound

Lightning and thunder occur simultaneously. However, there is a time lag between the thunder being heard and the lightning being seen by an observer, the lightning

---

Experiment 4–5
Connect a microphone to a cathode-ray oscilloscope. A microphone transforms the sound incident on it into electrical signals which vary in a similar manner to the sound intensity. The oscilloscope shows how the voltage produced by the microscope varies with time. Examine the sound produced by a loudspeaker. Try various frequencies. How does the trace compare with that when the signal generator terminals are connected directly to the oscilloscope? Examine the sounds produced by various musical instruments. Examine noise.

Experiment 4–6
Connect the microphone to the cathode-ray oscilloscope. Place the microphone so that it will pick up the sounds produced by two pieces of wood struck together and their echo returning from a reflector placed about 50 cm away (Fig. 4–3). The oscilloscope should

have its time scale, the x axis, adjusted so that time differences of the order of milliseconds can be determined. Bang the two pieces of wood together and observe on the oscilloscope the positions of the initial and the reflected pulse. Measure the time interval and

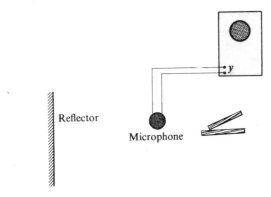

Fig. 4–3 Speed-of-sound experiment

being detected first. The time lag depends on the distance from the observer at which the thunder and lightning occur. This shows that sound travels more slowly than light. One method of measuring the speed of sound is to produce a short-duration sound and determine the time interval between the emission of the sound and its echo from a distant reflector.

(Experiment 4–6.) Sound in air at 0 °C travels at about 330 m/s; at 20 °C the speed is about 344 m/s. The speed of sound depends on the temperature. The speed depends on how long it takes a molecule to pass a vibratory motion onto the next molecule.

(Experiments 4–7, 4–8, and 4–9.) Sound requires a material medium for its propagation; thus it can travel through solids, liquids, and gases but not through a vacuum. The speed in water at 20 °C is 1,460 m/s, about four times faster than the speed in air. In steel at 20 °C the speed is 4,990 m/s, about fourteen times faster than in air.

Ultrasonic waves are used in industry to locate flaws in castings. The ultrasonic source, together with the detector, is placed in good contact with the casting. If there is a flaw present, the ultrasonic wave is reflected by it, and a signal is received at the detector sooner than the signal which travels all the way through to the rear surface of the material and back. Essentially the method is the same as that used to measure the speed of sound by its echo; in the present case two echoes are produced, one by reflection at the flaw and one by reflection at the rear surface of the casting.

Example 4–2. Ultrasonics are used in an experiment to determine the location of a flaw in a steel specimen. Two echoes are produced, one 1/3,000 s and the other 1/7,500 s after the initial pulse. What is the thickness of the specimen and where is the flaw located? The speed of sound in steel may be taken as 5,000 m/s.

The longest time interval is due to the full thickness

---

the distance over which the sound travelled, there and back. Hence determine the speed of sound in air.

For larger distances of 300 m or more an oscilloscope is not necessary—a stopwatch can be used.

Experiment 4–7
Place an electric bell under a bell-jar. Start the vacuum pump to evacuate the bell-jar. What happens to the sound heard outside the jar as the pressure inside is reduced? Does sound travel through a vacuum?

Experiment 4–8
Place a tuning fork against one end of a metal rod; place the other end against your ear. Try other materials. Does sound travel through solids?

Experiment 4–9
Produce a sound under water in a trough by banging two objects together. Can you hear the sound when you place your ear against the side of the vessel? Does sound travel through water?

Experiment 4–10
A metal rod or pipe at least 5 m long is required. One possibility is to use the gas or water piping in the laboratory. Connect two microphones to the oscilloscope and place them a large distance apart and in contact with the metal pipe. Strike the metal pipe sharply near one microphone and observe the pulses on the oscilloscope as each microphone detects the sound passing it. Measure the distance between the microphones and the difference in time; hence obtain a value for the speed of sound in the pipe material.

Experiment 4–11
Connect a small loudspeaker or an earphone to a signal generator. To produce a narrow beam of sound place the speaker at the end of a length of open cardboard

of the specimen being transversed twice by the pulse; thus the thickness is given by

$$\text{thickness} = \text{velocity} \times \tfrac{1}{2} \text{ (time of echo)}$$

$$\text{thickness} = 5{,}000 \times \frac{1}{6{,}000} = 0\!\cdot\!83 \text{ m}$$

$$\text{distance to the flaw} = 5{,}000 \times \frac{1}{15{,}000} = 0\!\cdot\!33 \text{ m}$$

Example 4–3. In the echo-sounding method of determining the depth of water under a ship, a pulse of sound is produced under the ship and the time taken for the sound to travel to the sea bed and back again is measured, that is the echo time measured. In a particular experiment the echo time was 2 and the velocity of sound was 1,500 m/s. What was the depth of water at that point?

$$\text{depth} = 1{,}500 \times 1 = 1{,}500 \text{ m}$$

(Experiment 4–10.)

4–4 Wave properties

(Experiments 4–11, 4–12, 4–13, and 4–14.) The experiments show that sound can be reflected and refracted. **Refraction** occurs when sound travels from one medium to another in which the speed of sound is different. Sound shows both **diffraction** and **interference.** The diffraction effects are most pronounced at long wavelengths, that is low frequencies.

Interference between two sounds of slightly different frequencies gives rise to the phenomenon of **beats.** The two waves interfere and produce a wave whose amplitude varies at a frequency equal to the difference in the two frequencies (Fig. 4–5).

(Experiment 4–15.)

Example 4–4. What is the beat frequency produced when two notes having frequencies of 500 and 510 Hz are sounded simultaneously?

---

tubing. Set the generator output at about 5 kHz and and detect the sound by means of a microphone placed at the end of another length of open tubing. The output from the microphone can be fed to an oscilloscope or a valve voltmeter.

Direct the beam of sound onto a reflector. Is sound reflected? Does the angle of incidence equal the angle of reflection?

Experiment 4–12

Place a loudspeaker and microphone facing each other and along the line between the two place a balloon filled with carbon dioxide. The balloon can be filled by placing a piece of solid carbon dioxide (dry ice) inside it when deflated. The output of the microphone can be detected by an oscilloscope or valve voltmeter. How is the detected sound affected by the presence of the balloon? Try varying the distance between microphone and the loudspeaker.

Experiment 4–13

Assemble the apparatus as in the diagram (Fig. 4–4). Direct the sound through a slit formed by two large screens. What happens if the microphone is moved about on the opposite side of the screens to the loudspeaker? Is sound diffracted? Try low and high frequencies. Explain the effect of changing the frequency?

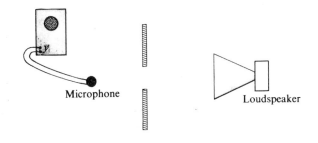

Fig. 4–4 Investigation of the diffraction of sound waves

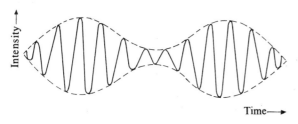

Fig. 4–5 *Amplitude variations when beats occur*

The beat frequency is the difference in the two frequencies, that is 510—500 Hz. Therefore the beat frequency = 10 Hz.

Summary

Sound is a longitudinal wave motion in a material medium. A longitudinal wave is one in which the particle displacement is along the same line as the direction of wave propagation. Reflection, refraction, diffraction, and interference effects can occur with sound. Sound is produced by an oscillation in a medium causing alternate compressions and rarefactions which travel through the medium. Sound generally travels faster in solids than in liquids and faster in liquids than in gases. The speed of sound in air at room temperature is about 340 m/s. Echoes produced by reflection of a sound pulse from a distant object can be used to determine its distance; this is the basis of echo sounding.

Problems

4–1 How does a longitudinal wave differ from a transverse wave?

*4–2 Is the diffraction of sound an everyday occurrence?

4–3 A sound of frequency 600 Hz travels through a medium at 300 m/s. What is the wavelength of the sound in that medium?

4–4 The higher the air temperature is, the greater the speed of sound. Explain what happens if close to a road surface

---

Experiment 4–14

Connect two loudspeakers to the same signal generator (Fig. 4–6) and place them about 1 m apart. How does the signal detected by the microphone vary as the microphone is moved along a line parallel to that of the speakers and about 2 or 3 m away? Try moving an ear along the same line (cover the other ear). Try both high and low frequencies. Do sound waves interfere?

Experiment 4–15

Connect two loudspeakers separately to two signal generators. Listen to the note produced when the two signal generators are set at the same frequency. What happens when the frequency of the signal generators is slowly changed?

Try connecting the two signal generators to the same oscilloscope. What happens to the trace when the frequency of one is changed?

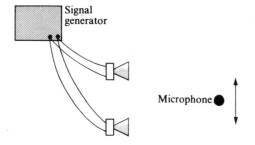

Fig. 4–6 *Investigation of interference with sound waves*

there is a layer of air at a higher temperature than the air further above.

4-5 Two loudspeakers situated 150 cm apart are connected to the same oscillator and emit a sound of frequency 500 Hz. Describe what occurs in the region of space in front of the loudspeakers.

4-6 What time interval will occur between the emission of a sound and its echo if the sound source is 100 m from a brick wall? Speed of sound = 340 m/s.

4-7 How are ultrasonic waves used for the detection of flaws in metal castings?

4-8 In an echo-sounding test carried out at sea the echo time was 4 s. If the speed of sound in sea water is 1,500 m/s, determine the distance between the source of the sound and the object giving the echo.

4-9 Ferromagnetic bodies alter in length under the action of a magnetic field. Thus, if a ferromagnetic rod is placed in a coil fed by alternating current, forced longitudinal vibrations are set up in the rod. At the resonant frequency of the rod this gives a convenient source of ultrasonic waves. A 6-kHz source is required—calculate the length of a suitable nickel rod which will generate this frequency. Speed of sound in nickel = 4,810 m/s.

5 Electric and Magnetic Fields

5–1 Electric charges

(Experiment 5–1.) Initially, no force is exerted between two balloons, A and B. When A and B are both rubbed with the same material, they are found to repel each other. We can say that the initially neutral balloons A and B have both gained the same kind of electric charge. If two balloons C and D are rubbed together, they attract each other. However, it will be found that A attracts C or D, and B attracts the one which is repelled by A. There are two kinds of electric charge, called **positive** and **negative electricity.**

Attractive forces occur between oppositely charged objects, while objects with the same charge exert **repulsive forces** on one another.

(Experiment 5–2.) An **electric current** is a movement or flow of electric charge and it does not matter whether it is obtained from a battery or a charged conductor. But what causes an electric charge to move? Water will move between two points if there is a difference in height between the water levels at the two points. In a somewhat similar manner we consider that electric charge will move between two points if there is a difference in electrical 'level' between them. To produce the difference in water levels originally, more work must have been done to lift the water to the higher level than to the lower level. Thus the water at the highest point will possess potential energy with respect to the lower point and will do work if allowed to descend. When two points have a difference in electrical 'level' we say that a **potential difference** exists between the two points and the point with the higher electrical 'level' is at a higher potential than the other point.

A difference of potential exists between two points if more work has been done to transport electric charge to one of the points than to the other. Potential difference between two points is defined as the work done in moving a unit positive charge from the point at lower potential to the point at higher potential:

$$\text{potential difference} = \frac{\text{work done}}{\text{positive charge moved}}$$

$$V = \frac{W}{q}$$

Units: potential difference   volts (V)
         work   joules (J)
         electric charge   coulombs (C)

Experiment 5–1

Suspend two balloons A and B by long nylon threads so that they almost touch each other. Rub them against your coat or sweater and allow them to hang freely What happens? Rub two balloons C and D together and suspend them like the first pair but about a metre away from A and B. What happens to the second pair? Now suspend one balloon, A say, from the first pair, close to one, D say, from the second pair. What happens?

Experiment 5–2

(a) Connect a battery in series with a resistance of about 1 MΩ and a spot galvanometer capable of detecting about a microampere of current. Observe the movement of the galvanometer spot with respect to the polarity of the battery.

(b) Charge the sphere of a van der Graaf generator. Connect the sphere to the galvanometer, the other side of which should be connected to the earth terminal of the generator. What happens?

Experiment 5–3

(a) Connect the sphere and the earth socket of the van der Graaf generator to two probes placed in an insulated container, a shallow glass dish. Pour a small quantity of oil into the dish to give a thin layer over the

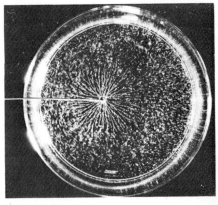

(a) A single charged rod.

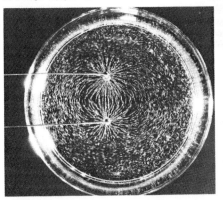

(b) Two rods with equal and opposite charges.

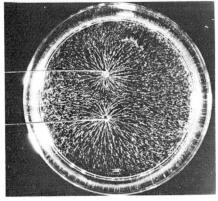

(c) Two rods with the same charge.

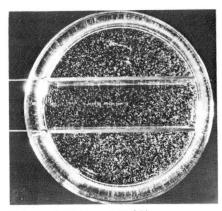

(d) Two parallel plates, no electric field.

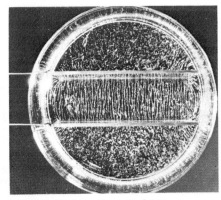

(e) Two parallel plates with opposite charges.

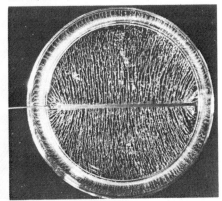

(f) A single charged metal plate.

Fig. 5–1 Electric force field patterns

5–2 Electric fields
(Experiment 5–3.) An **electric field** is that region of space where a force is experienced by another charged body placed there. The **electric field strength** at a point is defined in terms of the force experienced by a charge placed in the field:

$$\text{electric field strength} = \frac{\text{force exerted}}{\text{charge placed at the point}}$$

$$E = \frac{F}{q}$$

Units: electric field strength   newtons per coulomb
                                                (N/C)

                                                  volts per metre (V/m)

      charge   coulomb (C)

      force   newtons (N)

The direction in which the electric field acts at a point is taken as the direction in which a positive charge would have moved if placed at that point. The grass seed sprinkled on the oil surface will form a pattern of lines between and around the electrodes (Fig. 5–1). These lines are called **lines of force.** A line of force is the path that would be described by a small positive charge if free to move in the field.

Example 5–1. Determine the force on a charged oil drop when placed in an electric field of strength 10^4 V/m. The drop carries a charge of 10^{-16} C. If the mass of the drop is 0·001 g, calculate its acceleration.

$$F = Eq = 10^4 \times 10^{-16} = 10^{-12} \text{ N}$$
$$F = ma$$

Therefore

$$a = 10^{-12}/10^{-6} = 10^{-6} \text{ m/s}^2$$

Example 5–2. Determine the kinetic energy acquired by a charge of 10^{-16} C when accelerated through a potential difference of 1,000 V.

The energy acquired is equal to the work done on the charge by the electric field:

$$\text{energy} = Vq = 1,000 \times 10^{-16} = 10^{-13} \text{ J}$$

5–3 Magnetic fields
(Experiments 5–4 and 5–5.) Magnetic fields are produced when an electric current flows in a conductor. The effect with a coil is greater than that with a single wire, and in the case of a long coil the magnetic field is similar to that of a bar magnet. In the case of a bar magnet it is considered that the magnet contains a large

---

bottom. Thinly sprinkle some fine grass seed evenly over the oil surface. What happens when the generator is running?

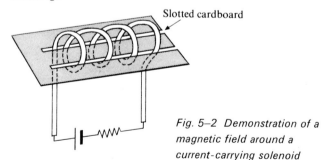

Fig. 5–2 Demonstration of a magnetic field around a current-carrying solenoid

(b) Replace the generator by an e.h.t. supply giving about 5 kV and repeat the experiment.

Experiment 5–4
Place a piece of paper over a bag magnet. Sprinkle iron filings evenly on the paper. What does the pattern of iron filings show?

Experiment 5–5
(a) Pass a piece of stiff wire perpendicularly through a hole in a horizontal piece of card. Sprinkle iron filings evenly on the card. Connect the wire to a high-current source. What does the pattern show?

number of very small magnets, each produced by the movement of charge. A measure of the strength of the field, the **magnetic flux density**, is given in terms of the force exerted on a current-carrying conductor placed in the field:

magnetic flux density B
$$= \frac{\text{force exerted on the conductor}}{\text{current} \times \text{length of conductor}}$$
or
$$B = \frac{F}{iL}$$

i is the current flowing in a length of wire L.

Units: flux density newtons per ampere metre
(N/A m)
webers per square metre
(Wb/m²)
force   newtons (N)
current   amperes (A)
length   metres (m)
(Experiment 5–6.)

5–4 Electromagnetic induction
An electric current flowing in a conductor gives rise to a magnetic field—can a magnetic field give rise to an electric current?

(Experiment 5–7.) An e.m.f. is induced in a circuit when it is in a varying magnetic field. The greater the rate of change of magnetic flux with time, the greater will be the induced e.m.f.

Summary
Unlike charges attract each other, like charges repel.

The rate of movement of charge is known as the **current.** For a current to flow between two points in a circuit there must be a **potential difference** between the points:

$$V = \frac{W}{q}$$

Grass seed sprinkled on the surface of oil round a charged object shows the existence of an **electric field.** The seeds set along the **lines of force** in the electric field:

$$E = \frac{F}{q}$$

In a similar manner iron filings sprinkled round a bar magnet or a current-carrying conductor show the existence of a **magnetic field**:

(b) Wind the wire into a small coil and insert a piece of slotted card into it, and repeat the experiment (Fig. 5–2). Compare the results with that of the bar magnet.

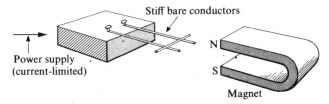

Fig. 5–3 *Apparatus for an investigation of the effect of a magnetic field on a current-carrying conductor*

Experiment 5–6
Connect two pieces of stiff parallel uninsulated wire to the terminals of the high-current d.c. supply (Fig. 5–3). Complete the circuit with a short piece of stiff wire which can slide along the other two. What happens to the movable wire when a magnet is brought near to it? What is the relationship between the directions of the current, the magnetic field, and the force on the conductor?

Experiment 5–7
Connect a small coil across the terminals of a sensitive galvanometer. Move a magnet in and out of the coil.

$$B = \frac{F}{iL}$$

A current gives rise to a magnetic field, and a changing magnetic field gives rise to an e.m.f. and hence a current in a conductor.

Problems

5–1 What evidence is there for the existence of two different types of electric charge, that is positive and negative?

5–2 If there is a potential difference of 5 V between two points, what work is done in moving 1 C of charge between the two points?

5–3 A charge of 2×10^{-4} C experiences a force of 20 N when placed in an electric field. What is the strength of the field?

*5–4 Can lines of electric force cross each other?

5–5 What is the kinetic energy acquired by a charge of $1 \cdot 6 \times 10^{-19}$ C when accelerated through a potential difference of 5 kV? If the mass of the charged object is 9×10^{-31} kg, calculate the velocity of the charge after the acceleration.

5–6 What do the iron-filing patterns round a current-carrying conductor or a permanent magnet show?

5–7 A wire of length 5 cm carries a current of 4 A and is placed in a magnetic field. What is the force experienced by the wire if the magnetic flux density at right angles to the wire is 1×10^{-3} Wb/m^2?

What happens when the magnet is moving and when it is stationary within the coil?

Move a single straight wire quickly through a strong magnetic field. What happens when the ends of the wire are connected to a sensitive galvanometer? How do the results depend on the speed at which the wire moves through the magnetic field or the magnetic field moves past the wire?

6 Heat

6–1 Forms of energy

(Experiment 6–1.) A body possesses energy if it is capable of doing work, that is of causing the point of application of a force to move through a distance. There are many forms of energy, e.g. mechanical energy, heat energy, chemical energy, electrical energy. The SI units of work are **joules**. As energy is purely stored-up work the same units are used. In heat a number of other units are also used. The **calorie** is the heat energy required to raise the temperature of 1 g of water by 1 degC. It is possible, however, to use the joule. In electricity one joule of energy is dissipated when a current of one ampere passes through a potential difference of one volt for one second.

energy per second = $IV \times J$

(Experiment 6–2.) Energy is always conserved. Although energy may change from one form to another, the total amount of energy will remain constant. This is known as the **principle of the conservation of energy**. Thus, when an object falls from a height, its potential energy is converted into kinetic energy. On impact the kinetic energy may be converted into heat. Regardless of the manner in which an experiment is performed, there is always one joule of mechanical or electrical energy for every joule of heat energy produced. If the process is reversed, one joule of heat energy will produce one joule of mechanical energy. Because of this conservation, energy is often considered to have only one unit—the joule. One joule of mechanical energy will produce one joule of heat energy or one joule of electrical energy.

6–2 Specific heat

The heat required to raise the temperature of 1 kg of

Experiment 6–1

(a) Lift a weight off the bench and let it fall back onto the bench. What energy changes are involved?

(b) Feel a piece of lead piping; then hammer it vigorously a number of times. Feel the lead again. What energy changes are involved?

(c) Short circuit a dry battery with a short length of wire. Feel the wire. What energy changes are involved?

(d) Strike a match. What energy changes are involved?

(e) Climb a flight of stairs. What energy changes are involved?

Experiment 6–2

Is energy conserved?

(a) Measure the temperature of a known mass of cold water in a cardboard cup; then measure the temperature of a known mass of hot water in another cardboard cup. Mix the two quantities of water and measure the resulting temperature. Is heat energy conserved?

(b) Measure the temperature of a known mass of cold water in a cardboard cup. Place a small 12-V immersion heater in the water and supply a steady current for a known time. Measure the potential difference across the heater, the current passing through it, and the temperature rise of the water. How many joules are needed to raise the temperature of 1 kg of water by 1 deg C?

(c) With the aid of an apparatus for the determination of the mechanical equivalent of heat (Fig. 6–1) determine the relationship between mechanical energy and heat energy. Essentially the apparatus consists of a copper cylinder around which a nylon tape is wound. One end of the tape carries a weight and the other end is fixed. Rotate the handle so that the force at the fixed end of the tape is zero. What force is exerted on the

water by 1 degC is 4,200 J. This is known as the **specific heat of water**, that is 4,200 J/kg degC.

Example 6–1. Calculate the amount of heat required to raise the temperature of 500 g of water by 5 degC.

The amount of heat required to raise the temperature of 1 kg of water through 1 degC is 4,200 J. The amount

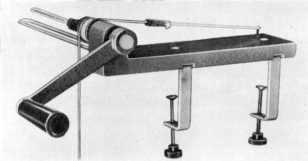

Fig. 6–1 Apparatus for determining the mechanical equivalent of heat

of heat required to raise the temperature of 500 g through 1 degC is 2,100 J. The amount of heat required to raise the temperature of 500 g through 5 degC is 105,000 J. Or

heat gained by water
$$= \text{mass} \times \text{specific heat} \times \text{rise in temperature}$$
$$= 0.5 \times 4,200 \times 5 = 10,500 \text{ J}$$

(Experiment 6–3.) The amount of heat required to raise the temperature of 1 kg of a material by 1 degC is known as the **specific heat** of that material. Typical values are 510 J/kg degC for stainless steel, 370 J/kg degC for brass, and 3,000 J/kg degC for sea water. The specific heat of a material depends on the temperature at which the measurement is made; the variation with temperature at room temperature is slight in most cases.

A method of determining the specific heat of a liquid is to use a continuous-flow calorimeter (Fig. 6–2). This consists of a central tube, through which the liquid

---

drum? Measure the diameter of the drum. What is the work done per revolution of the drum? Determine the temperature rise for a certain number of revolutions. What is producing the rise in temperature? Determine the heat required to raise the temperature of the copper cylinder by 1 degC and hence obtain a value for the heat produced in this experiment. What is the relationship between the heat produced and the mechanical energy?

The heat necessary to raise the temperature of the cylinder by 1 degC can be found by supplying a known amount of electrical energy to a cylinder via the coil embedded in it or by placing it in a cardboard cup containing warm water of known temperature and by determining the fall in temperature.

Experiment 6–3
Determine the specific heat of aluminium. Use a cylinder of aluminium drilled to accept an immersion heater

and a thermometer. Oil should be placed in the holes to ensure good thermal contact. Supply a known amount of electrical energy to the cylinder and measure the mass of the aluminium and the change in temperature.

Experiment 6–4
Use a continuous-flow calorimeter to determine the specific heat of water. Comment on the accuracy achieved and the factors limiting the accuracy.

Experiment 6–5
Use the continuous-flow calorimeter to determine a value for the specific heat of air at atmospheric pressure.

Experiment 6–6
For this experiment a special piece of apparatus known as a heat-of-combustion apparatus may be used or the apparatus can be made up with standard laboratory ware. The fuel is burnt in what is essentially a spirit

flows, containing an electrically heated wire. The central tube is surrounded by a vacuum jacket to reduce heat loss. The temperatures of the liquid entering and leaving the tube are measured—when steady conditions have been reached. The mass of liquid passing through the apparatus per second is determined by collecting it as it leaves the tube.

energy supplied per second by the heater $= IV \times J$

If θ_1 and θ_2 are the inlet and outlet temperatures, then

heat gained by the liquid per second $= ms(\theta_2 - \theta_1)$

where m is the mass of liquid passing through the apparatus per second, and s is the specific heat of the liquid. Some heat losses will occur and these can be represented by h.

$$IV = ms(\theta_2 - \theta_1) + h$$

Their effect can, however, be eliminated to a reasonable

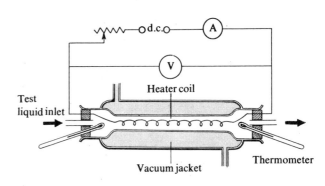

Fig. 6–2 Continuous-flow calorimeter

extent if the experiment is repeated for different values of I, V, and m but the same temperatures. (Experiment 6–4.)

Gases have specific heats; these specific heats are, however, complicated by the fact that gases can under-

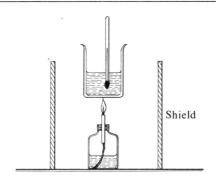

Fig. 6–3 Heat-of-combustion apparatus

lamp (Fig. 6–3). The heat produced by the fuel is determined by measuring the rise in temperature of a can of water. A metal can holding about 300 ml of water should be used. The entire apparatus should be shielded from draughts by sheets of asbestos or other suitable material.

The lamp should be weighed containing fuel; methanol or ethanol may be used. The weight of water in the can should be determined and its temperature noted immediately prior to the experiment. The water temperature should be measured again about 5 min after the lamp is lit; stirring the water is essential. The lamp should then be extinguished and the lamp plus fuel reweighed. How much heat was produced per unit mass of fuel burnt?

If time permits, try other fuels and compare the results. For solids the commercial apparatus is necessary. In this solids are allowed to burn in a stream of oxygen. The hot gaseous products then pass through a coil heat exchanger, immersed in water. Otherwise the procedure is as before.

Experiment 6–7

Heat some naphthalene in a test-tube until it melts. This is best done by dipping the test-tube into a beaker

go large changes in pressure and volume when heated. Two principal specific heats are used—the **specific heat at constant volume** and the **specific heat at constant pressure**. The specific heat at constant volume is the heat required to raise the temperature of 1 kg of the gas by 1 degC at constant volume. The specific heat at constant pressure is the heat required to raise the temperature of 1 kg of the gas by 1 degC at constant pressure. The two specific heats are not the same, e.g. air has the values 720 J/kg degC at constant volume and 1,010 J/kg degC at constant pressure. In the case of constant volume the heat energy supplied is used solely to increase the temperature of the gas, while at constant pressure the heat energy is used to raise the temperature and expand the gas. In the expansion, work is done, a force moves through a distance, and so energy is used.

(Experiment 6–5.)

Example 6–2.   An air heater consists of an electrically heated wire over which air is drawn. Calculate the power required for the heater if, for an air flow of 10 kg/min, the temperature is to be raised by 20 degC. The specific heat of air at constant pressure is 1,010 J/kg degC.

If W is the power, then the heat supplied per second is W joules.

$$\text{heat required per second} = \frac{10}{60} \times 1{,}010 \times 20 \text{ J}$$

If all the supplied energy is used to heat the air,

$$W = \frac{10}{60} \times 1{,}010 \times 20 = 337$$

Hence the power required is 337 W.

6–3 Chemical energy

When you strike a match, heat is produced by friction. This raises the chemical compound of the match head to its combustion temperature, after which heat is produced by a chemical change. Every substance can be

---

of hot water. Place a thermometer in the liquid naphthalene and observe how the temperature varies with time as the naphthalene cools and solidifies. Plot a graph of temperature against time. What is the significance of the graph?

Experiment 6–8
Determine the latent heat of fusion for ice. Place two funnels containing equal amounts of ice over beakers (Fig. 6–4). Place an electric immersion heater in each funnel but only connect one to an electrical supply. Supply a known amount of electrical energy to one of the heaters. Determine the mass of ice melted. Also determine the amount of ice melted in the funnel which was not heated by the immersion heater. Calculate the latent heat of fusion of ice. Why the arrangement with an unheated heater?

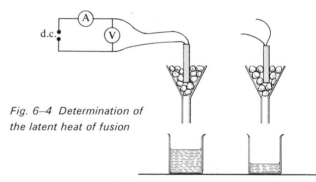

Fig. 6–4 Determination of the latent heat of fusion

Experiment 6–9
Place a number of marbles in a tray. Stick the marbles together with Vaseline. Shake the tray. Could this represent heat being supplied to a solid? Shake the tray vigorously until the marbles break apart. Could this represent melting? Energy is used to cause agitation of a solid when it is heated and also to break the bonds between the marbles.

considered to be a source of energy, and changes in its chemical composition may release some of that energy. The change may also be such that energy is taken in from the surroundings. Energy can in many cases be obtained by burning the substance in air. Materials which supply energy by this process are known as **fuels**. Coal and oil are examples.

(Experiment 6–6.)

6–4 Change of state

(Experiment 6–7.) During the change from solid to liquid and also liquid to solid no change in temperature occurs. In the case of the change from solid to liquid, heat is supplied without any change in temperature being produced—energy is necessary for the change. For the change from liquid to solid, heat is given out without any change in temperature occurring.

The amount of heat necessary to change 1 kg of solid to a liquid without changing the temperature is known as the **latent heat of fusion**. Thus, if heat energy Q melts a mass m, without any temperature change, the latent heat L is given by

$$L = Q/m$$

Example 6–3. Calculate the amount of heat required to melt 12 kg of ice. The ice is at 0 °C and the latent heat of fusion of ice is $3·34 \times 10^5$ J/kg.

$$\text{heat required} = 12 \times 3·34 \times 10^5 = 40 \times 10^5 \text{ J}$$

(Experiments 6–8 and 6–9.)

Energy is required to change a liquid into a vapour. The heat necessary to change 1 kg of liquid to vapour, without a change of temperature, is known as the **latent heat of vaporization**.

(Experiment 6–10.)

6–5 Vapour pressure

(Experiment 6–11.) Ether, or indeed any liquid, in

Now shake the tray so that the 'liquid' marbles fly out of the tray. Does the effect resemble that of a vapour?

Experiment 6–10

Heat some water in a beaker to boiling point with an immersion heater. Mark the level of the water on the side of the beaker and add further energy. The amount of energy added can be determined by measuring the current and voltage supplied to the heater and the time. When a measurable quantity of water has been vaporized mark the level. Determine the amount of water vaporized by adding water from a measuring cylinder to bring the level back to the mark again. Hence calculate the latent heat of vaporization of water.

Experiment 6–11

(a) Place a small amount of ether on the bench. What happens? How are conditions different from in the bottle?

(b) Fill a barometer tube with mercury and invert it in a trough of mercury. Because of the atmospheric pressure the resulting column of mercury will be about 76 cm high. What is above the mercury? Now introduce ether into the space above the mercury – take care not to introduce air. The ether can be introduced with a syringe. What happens to the mercury level? Offer an explanation. Introduce more ether. What happens? Add more ether until liquid ether remains on top of the mercury. What does the final mercury level indicate?

Experiment 6–12

Assemble the apparatus as shown in the diagram (Fig. 6–5). The pressure above the surface of the liquid in the flask can be reduced by means of a filter pump.

vapour form exerts a pressure on the mercury level and causes it to be depressed. This is known as the **vapour pressure**. If sufficient liquid is present, not all of it will evaporate and the resulting pressure is known as the **saturation vapour pressure**. In this condition the liquid is evaporating to form vapour and the vapour is condensing into a liquid at the same rate—the vapour pressure has thus reached a maximum. The vapour pressure depends on temperature.

Evaporation takes place from surfaces, but boiling takes place from within the body of a liquid. Bubbles of vapour are formed which move to the surface. Bubbles can only be formed if the vapour pressure is equal to, or greater than, the pressure on the surface of the liquid, that is the saturation vapour pressure is equal to the atmospheric pressure.

(Experiments 6–12, 6–13, and 6–14.) The temperature at which the change of state occurs depends on the vapour pressure, and a change in the pressure can be brought about by changes in the substance used and the addition of other substances. The external pressure also determines the vapour pressure at which the change can occur.

(Experiment 6–15.) For a liquid to change to a vapour energy is necessary, the latent heat of vaporization. When a liquid evaporates, the energy is supplied by the liquid itself and thus energy is abstracted from the liquid. This results in a drop in temperature.

Example 6–4.   In an industrial refrigerator, ammonia is vaporized in order to produce the low temperature. How much ammonia must be evaporated to freeze a kilogramme of water; the water may be assumed to be at $0\,°C$?

latent heat of fusion of water $= 3{\cdot}34 \times 10^5$ J/kg

latent heat of vaporization of ammonia
  $= 1{\cdot}34 \times 10^6$ J/kg

heat necessary to freeze 1 kg of water $= 3{\cdot}34 \times 10^5$ J

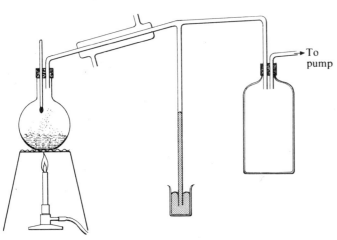

Fig. 6–5 Apparatus for the determination of the variation of the boiling point of water with pressure

The long tube, about 80 cm long, dipping into mercury enables the pressure in the apparatus to be determined. The condenser is to return the vapour back to the flask. Determine how the boiling point varies with pressure.

Experiment 6–13
Place some water and a thermometer in a flask and determine the temperature at which boiling occurs. Add salt to the water. Is there any effect on the boiling point? A thermometer should be used to give temperature readings both below and above the surface of the water.

Experiment 6–14
Determine the freezing point of water. This can be done by placing a test-tube containing water in a freezing mixture of ice and salt in a beaker. Now add salt or antifreeze to the water. What is the effect of this on the freezing point?

If m is the mass of ammonia that must be evaporated,

$$1\cdot34\times10^6\times m = 3\cdot34\times10^5$$
$$m = 0\cdot248\ \text{kg}$$

6–6 Gases and vapours

(Experiments 6–16, 6–17, and 6–18.) The results of experiments show that, for a gas like air, the pressure is inversely proportional to the volume at constant temperature:

$$P \propto 1/V$$

When a graph is plotted of the pressure against the reciprocal of the volume the graph appears to pass through the point $P = 0$ and $1/V = 0$. The relationship is therefore of the form

$$P = K\times\frac{1}{V}$$

where K is a constant.

The pressure and volume variations with temperature are

$$P \propto \theta \qquad \text{at constant volume}$$
$$V \propto \theta \qquad \text{at constant pressure}$$

where θ is the temperature in degrees Celsius. These graphs appear to pass through the zero pressure or volume axis at a temperature of $-273\ ^\circ$C. This does not happen because all gases liquefy and solidify before that temperature is reached. However, the idea is useful and $-273\ ^\circ$C is referred to as the **absolute zero of temperature**. Temperatures measured from this zero are said to be on the absolute or Kelvin temperature scale: $1\ ^\circ$C $= 1\ ^\circ$K. Thus $0\ ^\circ$C is $273\ ^\circ$K. This gives for our experimentally determined equations

$$P = K'T \qquad \text{and} \qquad V = K''T$$

where T is the temperature on the Kelvin scale, and K' and K'' are constants. When temperatures are reck-

---

Experiment 6–15
Place a small amount of ether on the back of your hand. What happens? Take care—ether is an anaesthetic, use only a very small amount.

Experiment 6–16
Determine how the volume of a gas, such as air, varies with pressure. A number of suitable types of apparatus are available. One in which the pressure is produced by a pump and measured with a Bourdon gauge is ideal (Fig. 6–6).

Experiment 6–17
Determine how the pressure of gas varies with temperature. This can be done by connecting a 250-ml flask to a Bourdon gauge (Fig. 6–7). The flask, which contains air, should be placed in a beaker containing ice and the pressure read. The temperature should then be raised by gently heating the flask in water.

Fig. 6–6

oned on the Kelvin scale, a doubling of temperature produces a doubling of pressure or volume whereas on the Celsius scale this is not the case. These three equations can be combined to give

$$PV/T = \text{a constant}$$

Example 6–5. How does the pressure of a constant volume of gas change when the temperature changes from 27 °C to 127 °C?

On the absolute scale of temperature 27 °C = 300 °K and 127 °C = 400 °K. Thus

$$\frac{P_1}{300} = \frac{P_2}{400}$$

$$\frac{P_1}{P_2} = \frac{3}{4}$$

(Experiment 6–19.) When both a vapour and a gas are present there will be pressure due to both. The experiment on the variation of boiling point with temperature has already shown how the saturation vapour pressure (s.v.p.) of water varies with temperature. Results over the range 0 °C to 100 °C are as follows.

s.v.p. (mmHg)	4·6	9·2	17·5	31·8	55·3	149	355	760
Temperature (°C)	0	10	20	30	40	60	80	100

Take your results from Experiment 6–19 and subtract the pressure due to the water vapour. The results should now be the same as those of the experiment in which only dry air was used. The conclusion from this is that the pressure of a number of gases or vapours is the sum of each constituent pressure taken separately. This is known as **Dalton's law.**

Example 6–6. The pressure in a container in which air and water are present is 770 mmHg at 20 °C. What will be the pressure at 60 °C if the conditions are still saturated?

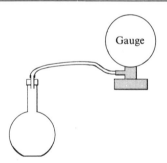

Fig. 6–7 Apparatus for the investigation of the effect of temperature on the pressure of a gas, the gas volume being constant

Experiment 6–18
Determine how the volume of a gas varies with temperature, the pressure being kept constant. This can be done using a 50-ml syringe (Fig. 6–8). A volume of dry air should be enclosed in the syringe. A small amount of phosphorous pentoxide placed inside the syringe will dry the air. The syringe should be air-tight but its plunger must be freely movable. Place the syringe in water and note how the volume of air inside the syringe varies with temperature.

An alternative method is to use a capillary tube sealed at one end and containing a short thread of mercury. The length of the air column between the mercury and the sealed end is then measured at different temperatures.

Experiment 6–19
In the previous experiments with air no water vapour should have been present. This can be absorbed by placing phosphorus pentoxide in the apparatus. Repeat Experiment 6–18 but remove the phosphorus pentoxide and introduce some water into the apparatus. With the capillary tube either the water can be in the sealed end of the tube or a water index can be used in place of the mercury. Take readings from 0 °C to as high a tempera-

The pressure at 20 °C is due to the air and the water vapour:

$$770 = P_w + P_a$$
$$770 = 17 \cdot 5 + P_a$$

Hence the air pressure is 752·5 mmHg. When the temperature changes the air pressure will change.

$$\frac{752 \cdot 5}{293} = \frac{P}{333}$$

Hence the air pressure at 60 °C is 855·1 mmHg. At 60 °C the vapour pressure due to the water will be 149 mmHg. Thus the combined pressure is 1,004 mmHg. Note that water vapour does not obey the gas laws.

The difference between a gas and a vapour is that a vapour can be liquefied by pressure alone, whereas a gas must be at a temperature below a certain critical temperature before pressure will cause it to liquefy.

s.v.p. (mmHg) for water	2,710	7,520
Temperature (°C)	140	180

ture as possible. Is there any difference between the results with water vapour present and those obtained without it?

Fig. 6–8 Apparatus for the investigation of the effect of temperature on the volume of a gas, the gas pressure being constant

Thus to liquefy steam at 180 °C a pressure of 7,520 mmHg must be applied. The critical temperature for water is 374 °C and above this temperature pressure alone will not cause liquefaction.

Summary

A **joule** of energy is dissipated when a current of one ampere passes through a potential difference of one volt for one second.

energy dissipated per second = IV = power

Energy is always conserved though it may change its form.

The **specific heat** of a substance is the amount of heat required to raise the temperature of 1 kg of the substance by 1 degC. Solids, liquids, and gases all have specific heats. In the case of gases two specific heats are generally quoted—one for conditions in which the volume is constant and the other when the pressure is constant.

Energy is needed to change the state of a substance from solid to liquid and from liquid to vapour. The energy needed to change the state of 1 kg of substance without change of temperature is known as the **latent heat**.

When a liquid evaporates, the vapour exerts a pressure known as the **vapour pressure.** If surplus liquid is present the pressure is known as the **saturation vapour pressure** (s.v.p.). The s.v.p. of a particular material depends on the temperature of that material. When the s.v.p. is equal to the external pressure, boiling occurs. Additives to a liquid or a solid can alter the temperature at which the change of state occurs.

A gas such as air obeys what are known as the **ideal gas laws**. These can be summarized in one equation

$$PV/T = \text{a constant}$$

In a mixture of gases or vapours the pressure exerted by the whole is the sum of the pressures exerted by the constituents.

Problems

6–1 A mass of 1 kg falls from a height of 1 m onto a bench. If the mass does not bounce, how much energy does it lose? In what form does it occur?

6–2 How much work is necessary to lift an object, of mass 6 kg, through a vertical distance of 200 cm?

6–3 An electric heater raises the temperature of 100 g of water in a cardboard cup through 50 degC in 5 min. How much energy is supplied to the water if none is assumed to be lost? What is the power of the heater? Specific heat of water = 4,200 J/kg degC.

6–4 An electric heater produces 0·5 kcal/s; what is its power in watts?

6–5 What quantity of heat is required to raise the temperature of 42 g of water by 10 degC? Specific heat of water = 4,200 J/kg degC.

*6–6 Explain what energy changes are involved in the following: (a) striking a match, (b) running up stairs, (c) a car moving along a road, and (d) a rocket burning its fuel.

6–7 How much energy is needed to change 30 g of ice into water at 0 °C? Latent heat of ice = $3·34 \times 10^5$ J/kg.

6–8 The latent heat of vaporization of water at 100 °C is $2·26 \times 10^6$ J/kg and the specific heat of water is 4,200 J/kg degC. How much heat energy is necessary to convert 100 g of water at 20 °C into steam at 100 °C?

6–9 Why is antifreeze added to the water in car radiators?

6–10 The heat combustion of a sample of petrol is $11·5 \times 10^3$ kcal/kg. How many joules of energy are released by the burning of 1 kg of petrol? What is the power of an engine if this amount of fuel is burnt in 15 s?

*6–11 Why do the brake drums of a car become hot when the brakes are applied?

6–12 Gas in a container has a pressure of 80 cmHg and a volume of 1 litre. What happens if the volume of the container is slowly reduced until it is 500 ml?

*6–13 Why are metereological balloons, which are used to take readings at high altitudes, only partially inflated when they leave the Earth's surface?

6–14 In the filling of a barometer tube with mercury a small amount of water finds its way into the tube. What effect will this have on the readings given by the barometer?

6–15 A container has both air and water present inside it. The pressure in the container is 76 cmHg at 20 °C. What will be the pressure at 80 °C? s.v.p. of water at 20 °C = 17·5 mmHg and at 80 °C = 355 mmHg.

6–16 Mercury used in a high-vacuum assembly is at a temperature of 25 °C. The saturation vapour pressure of mercury at that temperature is $1·7 \times 10^{-3}$ mmHg. Comment on the significance of these data.

7 The Measurement of Temperature

7–1 Temperature and temperature scales

Temperature can be defined as the hotness or coldness of a substance. If one substance is at a higher temperature than another, heat will flow from the high- to the low-temperature substance until the two reach the same temperature. Temperatures can be compared by utilizing properties which change with a change of temperature, e.g. the volume of a fluid, electrical resistance, the brightness of the light emitted by a glowing body. Temperature measurements are always comparisons, and, in order to make readings obtained by different observers agree, standard temperatures have to be specified—a temperature scale has to be defined.

(Experiments 7–1 and 7–2.)

In addition to specifying fixed temperature points we must subdivide the interval between them to give convenient units or degrees of temperature. The scale given by a thermistor, assuming a linear relationship between temperature and resistance, differs from that obtained with a gas thermometer where a linear relationship was assumed between pressure and temperature. In order to remove this difficulty the **International Scale of Temperature** is used, in which fixed points and the means of establishing temperatures between them are specified.

7–2 Measurement of temperature

(Experiment 7–3.) If the pressure of a gas at room temperature is extrapolated back to the zero-pressure point, a value of about $-273\,°C$ is obtained for the corresponding temperature and this is known as **absolute zero**.

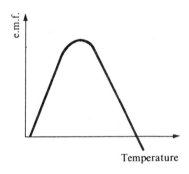

Fig. 7–1 Variation of thermoelectric e.m.f. with temperature

Experiment 7–1

A thermistor is a type of resistor which gives large changes in resistance with temperature. The resistance can be measured with an Avometer or with an ammeter and voltmeter and by use of Ohm's law.

Establish two fixed end points for the temperature scale. Two possible points are the freezing and boiling points of water. Call them 0 and 100 (use other numbers if you prefer). Measure the resistance of the thermistor in a mixture of ice and water and then in the steam immediately above boiling water. Also measure the resistance at room temperature. How can you establish a number for the room temperature? What assumption must you make?

Experiment 7–2

In this experiment establish your temperature scale by measuring the way the pressure of a gas, at constant volume, varies with temperature. The pressure in a flask of air can be measured by connecting it to a Bourdon gauge. Use the same fixed points as before and denote them by the same numbers. What is the room temperature on this scale? How does it compare with that obtained with the thermistor? Comment on the results.

(Experiment 7–4.) When the junction of two dissimilar metals is heated, a potential difference is produced between the two metals. This potential difference depends on the temperature of the junction (Fig. 7–1). If the other ends of the wires are connected to a galvanometer, they also form a thermo-junction whose temperature, and hence generated potential difference, depends on the temperature at the galvanometer end of the wires. In order to make accurate temperature measurements two thermocouples are joined together. One junction is attached to the body whose temperature is to be measured and the other is placed in a beaker of melting ice (at 0 °C). The combined potential difference then obtained—of the order of a few millivolts—indicates the temperature of the hot body against a constant (0 °C) reference (Fig. 7–2). A thermocouple has a low thermal capacity and thus does not significantly affect the temperature of an object when placed in contact with it. The gas thermometer has a high thermal capacity and does not measure the temperature over a small region like a thermocouple.

(Experiment 7–5.) The resistance of most metals increases with an increase in temperature. This forms the basis of an accurate method of determining temperature. The platinum resistance thermometer is one of the instruments used to specify the International Scale of Temperature. The resistance element consists of a specified length of platinum wire wound on a bobbin and placed in a protective sheath. In order to compensate for the variation in resistance (due to temperature variation) of the leads from the coil to the bridge, an identical pair of leads are placed alongside the first pair and connected to the opposite arm of the Wheatstone bridge. To a first approximation the resistance of a metal is a linear function of temperature:

$$R_t = R_0(1 + At)$$

where A is the temperature coefficient of resistance. For

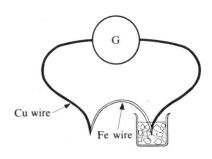

Fig. 7–2 Thermocouple circuit

Experiment 7–3

Repeat Experiment 7–2 in which the pressure of a gas is measured as a function of temperature. Calibrate the thermometer at the steam point 100 °C, the ice point 0 °C, and at temperatures below this. A mixture of solid carbon dioxide and methylated spirits gives a temperature of −72 °C. If liquid nitrogen, which boils at −196 °C, is available, try the experiment with this also. At what temperature would you expect the pressure of the air inside the flask to be zero? Explain the result obtained with the liquid nitrogen. What would you expect to happen as the temperature becomes progressively lower?

Experiment 7–4

Take about 40 cm of copper wire and a similar length of constantan or iron wire. Twist one end of the copper wire tightly round one end of the other wire, after both ends have been scraped clean, so that a junction is formed. Such a device is called a **thermocouple.** Connect the other ends of the wires to a sensitive galvanometer (microampere sensitivity). What happens when you warm the junction of the two wires by holding it in your hand? Calibrate the thermocouple against either a mercury-in-glass thermometer or the gas thermometer of the previous experiment.

accurate work a more general equation is used as the variation is not perfectly linear:

$$R_t = R_0(1 + At + Bt^2)$$

A is considerably greater than the constant B.

(Experiment 7–6.) Basically there are two forms of pyrometer: the **disappearing-filament** type, in which the current necessary for a filament to have the same brightness as the hot object whose temperature is to be measured is determined, and the **total radiation** pyrometer, in which the energy emitted by a unit area of, say, a furnace is determined by the use of a thermocouple of a photoconductive cell (Fig. 7–3).

Probably the most frequently used instrument for the measurement of temperature is the mercury-in-glass thermometer. This is usually employed to measure the temperature of liquids and gases, as thermal contact between the thermometer and a solid is not always easily achieved. The depth of immersion of the thermometer in the fluid affects the reading—the upper part of the stem may not be at the same temperature as the bulb. Thermometers should always be immersed to the same point on the stem as when they were calibrated: this point is marked by the makers.

(Experiment 7–7.)

Another form of liquid-in-glass thermometer

employs a Bourdon gauge to give an indication of a change in volume. The gauge is connected to a steel bulb by steel tubing and the whole is filled with mercury. When the mercury expands, the Bourdon gauge

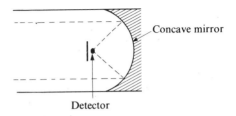

Fig. 7–3 Total radiation pyrometer

registers the volume change by partially uncoiling and moving a pointer across a scale. The Bourdon gauge works on the same principle as the toys consisting of a rolled-up paper tube which unrolls when blown into. The advantage of this form of thermometer is that the gauge may be at some distance from the bulb.

7–3 Industrial temperature measurements
The following are the range over which the various instruments are used and comments are given on the various limitations.

Experiment 7–5

The **thermistor** is a semiconductor resistance element and has a negative temperature coefficient of resistance, that is as the temperature increases the resistance decreases.

temperature coefficient of electrical resistance

$$= \frac{R_t - R_0}{R_0 t}$$

where R_t is the resistance at t °C, and R_0 is the resistance at 0 °C.

Determine how the resistance of (a) a thermistor, and (b) a small coil of Nichrome or fine copper wire

varies with temperature. Calibrate them against either your standard gas thermometer or a mercury-in-glass thermometer. The resistance of the metal wire coil

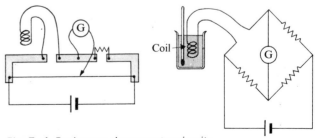

Fig. 7–4 Resistance thermometer circuits

Liquid-in-glass thermometer: −200 to 500 °C cannot be used for distant reading

Liquid-in-steel: 0 to 600 °C suitable for distant reading

Electrical resistance: −240 to 600 °C suitable for distant reading, accurate

Base-metal thermocouples: −200 to 1,100 °C suitable for distant reading

Rare-metal thermocouples: 0 to 1,450 °C suitable for distant reading

Disappearing-filament pyrometer: 700 °C upwards

Total radiation pyrometer: 500 °C upwards

In many industrial processes a record of temperature variation with time is required. This is obtained by the thermometer giving an electrical signal related to the temperature which actuates what is essentially a moving-coil galvanometer. The pointer of the instrument produces an ink trace across a moving strip of paper. These recorders are particularly used with thermocouples, which give a current output, and resistance thermometers, which when placed in an arm of a Wheatstone bridge give an out-of-balance current related to the resistance change.

7–4 International Scale of Temperature

The primary fixed points are as follows.

Boiling point of liquid oxygen	−182·97°
Ice point (triple point)	0·01°
Boiling point of water	100°
Boiling point of sulphur	444·6°
Freezing point of silver	960·8°
Freezing point of gold	1,063°

All these points are rigorously defined so that the same conditions and hence the same temperatures can always be obtained. For example, the 0·01° point is defined as the temperature of equilibrium between ice, water, and water vapour **(triple point of water)**:

$$t = 0·01 - (0·7 \times 10^{-6}\,°/\text{mm})\,h$$

h is the depth in millimetres below the liquid–vapour surface at which the temperature is determined. The boiling points are defined in terms of the pressure at which they are measured, since pressure alters the boiling point.

should be measured with a metre bridge or other form of Wheatstone bridge (Fig. 7–4).

Experiment 7–6

Connect a low-voltage lamp to a d.c. supply via a rheostat and a milliammeter (Fig. 7–5). As the current is increased, what happens to the brightness of the light emitted by the lamp filament?

The values of the current giving the different brightnesses can be used as a measure of the temperature of the filament. When the lighted lamp filament is observed against a furnace as background and the current varied until the two are the same brightness, the

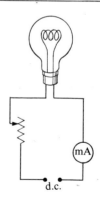

Fig. 7–5 Disappearing-filament pyrometer

In order to specify points between the fixed points the method of measuring the temperature is specified. For example from $0°$ to $630·5°$ the temperature is defined by the formula

$$R_t = R_0(1 + At + Bt^2)$$

where R_t is the electrical resistance of a platinum wire at temperature t and R_0 its resistance at $0°$. A and B are constants which are determined by measurements of the resistance at the boiling point of water, the boiling point of sulphur, or the freezing point of zinc. The annealed platinum must be of such purity that R_{100}/R_0 is not less than $1·392$.

From $-182·97°$ to $0°$ a platinum resistance thermometer is used, with a different equation from the above, while from $630·5°$ to $1,063°$ a thermocouple is used. Above $1,063°$ the temperature is defined in terms of the brightness of the light emitted by the hot substance.

Temperatures measured on this scale, the **International Practical Scale of Temperature,** are expressed in **degrees Celsius,** designated by $°C$. This scale depends on the properties of various thermometers. A **gas scale** or **absolute scale** has been devised and depends on the properties of gases. This gives much the same scale as the International Practical Scale. A scale which does not depend on the properties of any material was devised by

Kelvin in 1848 and is known as the **thermodynamic scale**. This scale was devised in terms of energy, but experimental difficulties in its realization led in 1927 to the designation and adoption of the International Scale (revised in 1948 and amended in 1960). The thermodynamic scale is the fundamental scale against which the practical scale is referred. The size of the degree has been fixed to give the thermodynamic temperature of the triple point of water as exactly $273·16°K$. Thermodynamic temperatures are designated as **degrees Kelvin.** The degree Kelvin is much the same as the degree Celsius and temperatures on the International Scale are basically the same as those on the thermodynamic scale minus $273·16°$.

Summary

The **International Practical Scale of Temperature** is based on a number of fixed temperature points, and a specification of the thermometers and formulae necessary to determine temperatures between the fixed points. Any physical property which changes with temperature can be used as the basis of a thermometer. The properties most used are the change of resistance with temperature, the change of volume with temperature, the change of pressure with temperature, the production

filament merges into the background, and they are then at the same temperature. The filament current is therefore a measure of the temperature of the furnace. The instrument is known as a **disappearing-filament pyrometer.** A colour filter (red) is included in the instrument so that the brightness comparison is made at one colour.

The pyrometer can be calibrated by heating various materials of known melting points in a small metal cup and by matching the brightness of the cup with that of the filament when the materials melt. Possible materials are sodium chloride 801 °C, potassium chloride 772 °C, and potassium sulphate 1,069 °C.

Experiment 7–7
Measure the temperature of boiling water in a beaker with the same thermometer immersed to different depths. How do the readings vary?

of a potential difference whose magnitude depends on temperature, and a change in the brightness of the light emitted by a hot object.

Problems

7–1 What type of thermometer would be suitable for the following measurements: (a) the temperature of a liquid (about 50 °C), (b) the surface temperature of a metal sheet (about 150 °C), (c) the temperature of a furnace (about 1,200 °C), (d) a temperature difference of about 0·01 deg C between two solids, (e) the temperature of a solid (about 60 °C)?

7–2 Why is it necessary to specify an international temperature scale?

7–3 How would you use a thermistor to measure the temperature of a hot liquid (about 150 °C) if the thermistor needs calibrating?

7–4 Why are melting and boiling points used as the fixed points on a temperature scale?

8 Optics

8-1 Reflection

(Experiments 8–1 and 8–2.) Light apparently travels in straight lines and can be reflected from mirrors. The angle of incidence is equal to the angle of reflection.

(Experiment 8–3.) Consider a beam of light incident on a plane mirror at an angle of incidence i. The angle of reflection is r.

$$i = r$$

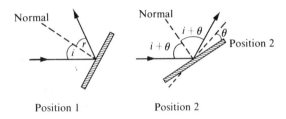

Fig. 8–1 The effect of rotating a mirror on the angle of incidence

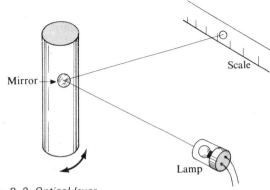

Fig. 8–2 Optical lever

If the mirror is rotated through an angle θ (Fig. 8–1), then the angle of incidence becomes $i + \theta$. The angle of reflection is therefore $i + \theta$. Before the rotation the angle between the incident and the reflected rays was $2i$. After the rotation the angle is $2(i + \theta)$. Therefore rotating the mirror by an angle has moved the reflection beam through an angle of 2θ.

Experiment 8–1

In a darkened room direct light from a powerful lamp, such as a tungsten iodide lamp, through a slit about 2 cm wide, onto a screen. Examine the shadow cast on the screen by an object placed between it and the slit. Does light travel in straight lines? Try various objects and examine the shadows. Blow smoke in the path of the light.

Experiment 8–2

Direct the light from a 24-W lamp through a slit and onto a plane mirror. The position of the slit should be altered until a narrow beam of light is produced and if necessary two slits should be used. Mark on a piece of paper the positions of the ray before and after reflection. Also mark the position of the mirror.

At the point at which the beam meets the mirror draw a line normal to the mirror (Fig. 8–3). Measure the angles of incidence and reflection, that is the angles between the normal and the incident, and the normal and the reflected beams. How do the angles compare? Repeat the experiment for different angles of incidence.

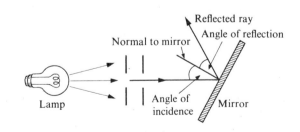

Fig. 8–3 Experiment 8–2

The reflection of a light beam from a mirror is used in what is known as an **optical lever**. This is used to magnify small angular rotations. A small plane mirror is fixed to the surface of the rotating object (Fig. 8–2). A beam of light is directed onto the mirror and the reflected beam allowed to fall on a scale. If the scale is a metre or more from the mirror, quite small angular movements of the mirror can be detected. Consider a mirror attached to an object rotating about an axis a distance r from the scale. If the distance of the scale to the axis of rotation is R, then the movement of the beam along the scale is $2R/r$ times the movement of the mirror (Fig. 8–4).

(Experiment 8–4.)

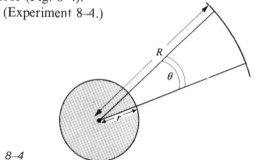

Fig. 8–4

(Experiment 8–5.) The paths of the reflected light beams with the plane mirror, in Experiment 8–5, will all appear to come from behind the mirror and, if they were produced back, they would appear to come from an image (Fig. 8–5) as far behind the mirror as the

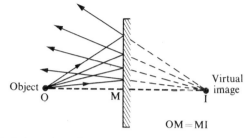

Fig. 8–5 Location of the image produced by a plane mirror

object is in front. This type of image is called a **virtual image**. With the concave mirror it is possible for the reflected rays to form a **real image** located in front of the mirror (Fig. 8–6).

(Experiment 8–6.) When the object is at infinity, that is the rays of light arriving at the mirror are parallel, the position of the image for a spherical mirror is known

---

Experiment 8–3
Direct a beam of light onto a plane mirror. Mark the positions of the mirror and the incident and the reflected beams. Rotate the mirror, through 10°, about the vertical line through the position where the beam meets the mirror. Mark the position of the new reflected ray. What is the relationship between the rotation of the mirror and the movement of the reflected beam? Try further rotations.

Experiment 8–4
Attach a small piece of plane mirror to the curved surface of a rod a few millimetres in diameter. Arrange the rod vertically and clamped at one end. The other end should be capable of being twisted. Direct a beam of

light onto the mirror and receive the reflected beam on a scale placed a few metres away. Twist the rod and observe the beam on the scale.

Experiment 8–5
Place a metal comb in front of a lamp in order to produce a number of beams of light. Direct the beams onto a curved mirror. What happens on reflection?

Replace the curved mirror by a plane mirror and compare the results. Mark the positions of the beams for both mirrors.

Experiment 8–6
Use a line-filament bulb as an object and place it in front of a concave spherical mirror. It should be placed

as the **focus.** Any incident ray of light parallel to the axis of the mirror will pass, or appear to pass, through the focus after reflection in the mirror.

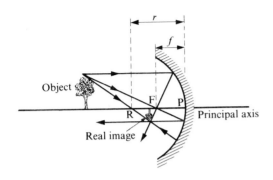

Fig. 8–6 Reflection of light by a concave mirror

(Experiment 8–7.) An object can be considered as a source of rays of light. These will travel in all directions from the object. Some will be parallel to the axis (Fig. 8–6) and pass through the focus after reflection; some will pass through the focus and emerge parallel to the axis after reflection. By drawing such rays it is possible

to determine the position of the image. When the image was produced on the screen, it was inverted; for some object positions, however, no image could be located by means of the screen, and on looking into the mirror an erect image will have been seen. As with the plane mirror, this image is behind the mirror and the light only appears to come from it. This is therefore a virtual image, the one produced on the screen a real image. When the object is between a concave mirror and its focus, then the image is virtual.

(Experiment 8–8.)

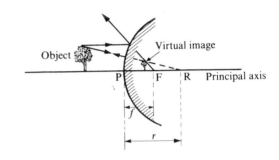

Fig. 8–7 Reflection of light by a convex mirror

slightly off the axis so that a small screen, a postcard, can be placed to receive the reflected light. Place the object at different distances from the mirror and locate the images with the card. What happens as the object distance, starting from about a metre, is reduced?

Experiment 8–7
Place a curved mirror on a piece of paper and direct a narrow parallel beam of light along the axis towards the mirror. What happens? Now direct a beam of light along a line parallel to the axis. What happens? Try other lines parallel to the axis. Mark the position of the focus, that is the point where the reflected beams intersect.

Direct a beam of light at an angle to the axis and passing through the focus. What happens? Try other similar rays.

Experiment 8–8
Use the metal comb and the lamp to give a number of beams of light and direct them onto a convex mirror. If the reflected rays diverge, extrapolate back to find the point from which they appear to come. Also look at images in the mirror. Where are the images for different object positions? Draw ray diagrams for object positions both between the focus and the mirror and on the far side of the focus from the mirror, and compare the results with those obtained in practice.

Experiment 8–9
Direct a single beam of light onto a rectangular block of glass. What happens to the beam? Is there any relationship between the angle of incidence and the

8–2 Refraction

(Experiment 8–9.) When a beam of light passes from one medium to another, 'bending' of the beam occurs, that is the direction of the light alters. This is known as **refraction**. The ratio of the sine of the angle of incidence to the sine of the angle of refraction is a constant for the two media concerned, and is known as the **refractive index**:

$$\text{refractive index} = \frac{\sin i}{\sin r}$$

(Experiment 8–10.) When light passes from a dense to a less dense medium, the angle of refraction in the less dense medium is greater than the angle of incidence. At a particular angle of incidence the angle of refraction can become 90°, that is the emergent beam travels along the interface of the two media. The angle of incidence at which this occurs is known as the **critical angle**. At angles of incidence greater than the critical angle, **total internal reflection** occurs, that is there is no refracted beam.

Example 8–1.   The refractive index for light passing

from glass to air is 0·6. What is the critical angle?

$$0 \cdot 6 = \frac{\sin i}{\sin 90°}$$
$$i = 37°$$

(Experiments 8–11 and 8–12.) Consider a ray of light incident on a glass-to-air boundary, refractive index 0·6. If the angle of incidence in the glass is 30°, then the angle of refraction can be calculated to be 56°.

$$0 \cdot 6 = \frac{\sin 30°}{\sin r}$$

If now we consider a ray incident on an air-to-glass boundary at an angle of incidence of 56°, it is found that the angle of refraction in the glass is 30°.

$$\text{refractive index for air to glass} = \frac{\sin 56°}{\sin 30°} = 1 \cdot 66$$

The refractive index in going from air to glass is the reciprocal of the refractive index in going from glass to air.

Example 8–2.   The refractive index for light crossing an

angle between the normal to the glass surface and the ray inside the glass?

Experiment 8–10
Take a semicircular block of glass and place it on a sheet of paper. Direct a beam of light towards the circular face and along a radius. Use different angles between the normal to the base of the block and the incident beam. What happens as the angle changes?

Also try a semicircular plastic cheese container filled with water.

Experiment 8–11
Either use the result of Experiment 8–10 or repeat the measurements to obtain a value for the critical angle and hence the refractive index. Place a small amount of

liquid between a glass slide and the base of the block. Obtain a value for the critical angle. What does the result represent?

Experiment 8–12
Is the path of a beam of light from one medium to another perfectly reversible? Thus, in the experiment with the semicircular block of glass and the beam of light passing through it from the curved to the flat side, would it matter if the ray started on the base or the curved side—would the path be the same? Try it.

Experiment 8–13
Direct a beam of light onto the face of a 60° prism. Try different angles of incidence. What happens? Try a 45° prism.

air-to-water boundary is $\frac{4}{3}$, what is the refractive index for light passing from water to air?

refractive index for water to air = $\frac{3}{4}$

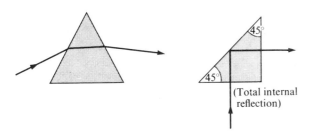

Fig. 8–8 Deviation of light by prisms

(Experiment 8–13.) A prism can be used to produce a greater deviation of a ray of light than a single refraction (Fig. 8–8). With the 45° prism internal reflection will be found to occur for the light incident on the hypotenuse side. Prisms are used in binoculars to bend the light path several times and hence shorten the distance between the object lens and the eyepiece, and they also provide a final erect image.

8–3 Lenses

(Experiment 8–14.) Both concave and convex lenses have focal points, that is points to which parallel rays converge or appear to converge after refraction by the lens. A convex lens produces real images when the object is outside the focal length and virtual images when it is inside. A concave lens produces virtual images for all object positions.

(Experiments 8–15 and 8–16.) Rays of light parallel to the axis pass, or appear to pass, through the focus after refraction. A ray through the focus is parallel to the axis after refraction. A ray through the centre of the lens is undeviated after refraction.

Essentially a **telescope** consists of a converging lens of long focal length which produces an image of a distant object; the image in turn acts as the object for a lens of short focal length, the eyepiece. This arrangement gives an inverted final image and, if the telescope is for ordinary terrestrial use, some means of erecting the final image must be employed.

In the **microscope** the object lens is of short focal length and produces an image of an object situated just outside the focal length. This image serves as an object

---

Experiment 8–14
Place the metal comb in front of the light source and produce a fan of rays. Allow them to fall on a cylindrical lens. Use plano-convex and plano-concave lenses of differing curvatures. What happens with the various lenses? Try moving the lenses to different distances from the light source. What happens?

Experiment 8–15
Direct a narrow parallel beam of light onto a convex lens. What happens to a beam parallel to the axis? What happens to a beam inclined to the axis and passing through the focus? What happens to a beam inclined to the axis and passing through the centre of the lens? Repeat the experiment with a concave lens.

Experiment 8–16
Look through convex and concave lenses at objects situated at different distances from the lenses. Compare your observations with the results of Experiment 8–14. Use the properties of rays found in Experiment 8–15 to sketch the ray diagrams necessary to explain your observations.

Experiment 8–17
Construct a telescope and a microscope.

Experiment 8–18
Place a metal comb in front of the light source, and with the aid of a convex lens produce an image on a screen. Now place a 60° prism near the lens and in the path of

for the eyepiece, another lens of short focal length. (Experiment 8–17.)

8–4 Dispersion

(Experiment 8–18.) White light on passing through a prism is dispersed into its constituent colours. As the different colours are refracted different amounts by the prism, this would suggest that the refractive index of glass depends on the colour of the light. The previous experiments on refraction should have shown this to some extent in that the refracted light would be seen to have coloured edges.

(Experiment 8–19.) **Chromatic aberration** is the name given to the defect of lenses in which the image appears coloured. In the case of white light no one image position exists, since there is an image for each constituent colour. A lens can be considered to be made up of a number of prisms (Fig. 8–9). It is possible almost to eliminate chromatic aberration by utilizing the fact that both convex and concave lenses deviate blue light more than red. By combining a convex lens and a concave lens it is possible to produce an **achromatic** lens which gives virtually no chromatic aberration. Two different types of glass must be used for the lenses as deviation of the light is still required—it must act as a lens. If the same glass were used, the resulting achromatic combination would be simply a block of glass.

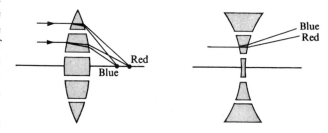

Colours not all brought to the same focus

Fig. 8–9 Lenses considered as being constructed of prisms

8–5 The nature of light

Light may be reflected or refracted—what else shows reflection and refraction?

(Experiment 8–20.)

Reflection and refraction can be shown by both particles and waves. A change in speed of both wave

---

the light. What is now the appearance of the image? Rotate the prism.

Cut a hole in the screen so that just one colour can pass through. Now with a second lens and prism produce another image on another screen of the light emerging from the hole. Is this property of light splitting up into colours something which happens with every beam of light, regardless of colour? What do the experiments show?

Experiment 8–19

Focus light from a distant lamp onto a screen with a convex lens. Move the screen through the focused position. How does the image change? Put a blue filter in front of the lamp and find the image position. Repeat with a red filter. Are the image positions the same? Explain what is happening.

Experiment 8–20

(a) Roll a marble towards a rigid plane vertical surface, a clamped glass block, and determine whether the angle of incidence is equal to the angle of reflection.

(b) With a ripple tank direct a pulse of waves towards a plane barrier. Is the angle of incidence equal to the angle of reflection? A pulse of plane waves can be produced by rocking a rod in the water.

(c) Roll a marble towards a curved rigid surface. Try aiming the marble at different points on the surface. Compare the results with those of light reflected at a curved mirror.

and particle are needed for refraction to occur. On this evidence light could be either a particle or a wave motion.

(Experiment 8–21.) When waves meet interference occurs, that is the waves interact and can give rise to varying degrees of disturbance from zero to a maximum at the points concerned. When two particles meet, no interaction occurs. Thus in front of the two sources of particles no particular pattern predominates, while in front of the two sources of waves a pattern arises owing to interference between the waves. With light a pattern similar to that occurring with water waves is found. The conclusion would seem to be that light is a wave motion.

Summary

Light undergoes **reflection** and **refraction.** With reflection the angle of incidence is equal to the angle of reflection. With refraction the sine of the angle of incidence divided by the sine of the angle of refraction is a constant determined by the two media concerned and known as the **refractive index.** Both particles and waves show reflection and refraction; however, only light and waves show **interference.** Thus light would appear to be a wave motion.

Reflection by a plane mirror produces **virtual images,** images apparently behind the mirror. The **focus** is the point to which parallel rays of light converge or from which they appear to diverge after reflection. With concave mirrors **real images** are produced when the object is outside the focus and **virtual images** when it is inside. With convex mirrors all the images are virtual, regardless of object position. Convex lenses produce real images with the object outside the focus and virtual images when it is inside.

At refraction white light splits up into its constituent colours. This splitting up or **dispersion** is particularly noticeable with a prism. Lenses, however, cause dispersion and this results in a coloured image when white light is used, as the focal length depends on the colour of the light.

Problems

8–1 A mirror is rotated through an angle of 10°. By how much is the reflected ray rotated if the direction of the incident ray remains unchanged?

(d) With a ripple tank direct a pulse of waves towards a curved barrier. Compare the results with those of light reflected at a curved surface.

(e) Roll a heavy ball bearing down a launching ramp, across a flat horizontal surface, down a slope to speed it up, and then along another flat horizontal surface. Direct the ball bearing at different angles of incidence onto the change in slope line (Fig. 8–10). What happens? The path of the ball bearing can be marked by allowing the ball bearing to run on a sheet of carbon paper which rests on a sheet of white paper.

(f) With a ripple tank direct a continuous plane wave onto a discontinuity produced by immersing a piece of Perspex or glass beneath the surface of the water. This produces a change in depth. For the best results the water over the sheet should be as shallow as

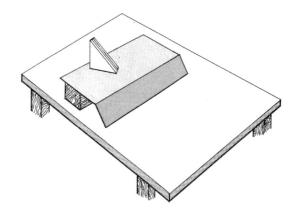

Fig. 8–10 Apparatus for the investigation of the 'refraction' of particles

8–2 Explain how you would use a mirror to measure the twist developed in a.5-cm cylinder when it is subjected to a couple.

8–3 Why are convex mirrors used as driving mirrors on cars?

8–4 Draw a ray diagram showing the location of the image for an object placed outside the focus of a concave mirror.

8–5 What determines the field of view seen by a car driver using a plane mirror located on the inside of the car windscreen?

8–6 An object is placed 5 cm in front of a concave mirror of focal length 10 cm. Describe the image and state its position.

8–7 By means of a ray diagram determine how big the image of the Sun will be for a telescope with an 18-m focal length concave mirror. The Sun is 1.5×10^{11} m distant from the Earth and has a diameter of 1.4×10^9 m.

8–8 A ray of light incident at an angle of 60° is refracted at an angle of 45°. What is the refractive index between the two media concerned?

8–9 What conditions are necessary for total internal reflection to occur?

*8–10 A solid has the same refractive index as a liquid. What will be seen when the solid is immersed in the liquid?

8–11 Why does a prism produce a greater deviation of light than a plane block of glass?

8–12 Draw a ray diagram for a telescope, consisting of a long-focus objective and a short-focus eyepiece lens.

8–13 By means of a ray diagram locate the image of an object situated 20 cm from a convex lens of focal length 15 cm.

8–14 Explain with the aid of a diagram how a slide projector works.

*8–15 Is white light made up of different coloured components or does a prism when acted on by light produce colours? How could you demonstrate the validity of your answer?

8–16 What is chromatic aberration and how can it be reduced for a lens used in a camera?

*8–17 Would it be possible to produce an analogue of the behaviour of light on passing through a lens by the use of ball bearings and different slopes?

---

possible. Try rectangular, prism-shaped, and lens-shaped pieces of glass or Perspex.

Experiment 8–21

(a) Use two launching ramps and direct the marbles from each ramp to give collisions. If light was particulate and we had two sources, what would be the result in the region where the light from the two sources met?

(b) Use two vibrators dipping into a ripple tank to give two sources of waves side by side. What happens where the waves meet? Use a stroboscope to 'freeze' the motion.

(c) Use the special ruling apparatus to rule two slits very close together on an Aquadag-coated microscope slide. A fine needle should be used to cut

the Aquadag. Hold the double slits about a metre away from a line-filament lamp in a darkened room. The filament should be parallel to the slits. Allow the light emerging from the slits to fall on a screen placed about a metre away. What can be seen in the region where the light coming through the two slits overlaps. Compare these observations with those of Parts (a) and (b).

9 Acoustics

9–1 The properties of sound

Sound is propagated in a medium by means of longitudinal waves. In common with all forms of wave motion, sound can be **reflected, refracted,** and **diffracted,** and shows **interference** effects (see Chapter 4). Sound is produced by vibrations and is propagated through a medium by a pressure wave, that is a series of compressions and rarefactions. Sound can travel through all material media, solids, liquids, and gases.

When a sound wave of a definite frequency impinges on the ear we say that a musical note is heard. The note is said to have a certain **pitch.** This is in fact the **frequency** and is described by a series of letters on a **musical scale.** For example, middle C is a frequency of 256 Hz and D is 288 Hz. Top C is 512 Hz. If the frequency is doubled from middle C to top C, a change of pitch of one octave has occurred. Many notes emitted by musical instruments are not notes of single frequencies—there is a difference between the middle C of a piano and that of a guitar or the human voice. The different notes are said to have different qualities.

(Experiment 9–1.)

Work must be done to produce sound, that is cause an object to vibrate and produce a pressure disturbance in a medium. Sound is thus a means of transmitting energy through a medium. The rate at which work is done is known as the power. **Sound intensity** is defined as the sound power passing through unit area, that is the amount of energy passing per second through unit area.

Units:	work	joules (J)
	power	joules per second (J/s)
		watts (W)
	intensity	watts per square metre (W/m^2)

(Experiment 9–2.) The ear is not equally sensitive to all frequencies, and thus the intensity necessary for a sound to be heard, the **threshold of audibility,** varies with the frequency, Sounds of the same intensity but different frequencies will, therefore, have different loudness (Fig. 9–1). For a given frequency the **loudness** of a sound is roughly proportional to the logarithm of intensity. Loudness is a difficult quantity to estimate as it depends very much on the observer concerned. All that is definite is that it is possible to tell when two sounds are of equal loudness. The lowest intensity that

---

Experiment 9–1

Connect a microphone to an oscilloscope. If the microphone has a low impedance, a transformer should be used to match the impedance of the microphone to the high impedance of the oscilloscope. Examine the waveforms produced by different musical instruments sounding the same pitch note.

Experiment 9–2

Connect a loudspeaker to a signal generator and measure the current to the speaker and the potential difference across it. Determine the power that must be supplied to the loudspeaker at different frequencies for the sound to be just heard. As the frequency region to which the ear is sensitive is large, plot your results as a graph of the logarithm of power against the logarithm of the frequency. Comment on the results.

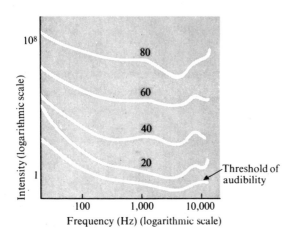

Threshold of audibility

Fig. 9–1 Curves of equal loudness (at 1,000 Hz, intensities are indicated in decibels above threshold,

can be heard by the human ear is about 10^{-12} W/m^2, while the highest is about 1 W/m^2. Because of this wide range of intensities a logarithmic scale is used and the intensity level I of a sound is defined by

$$n = \lg\left(\frac{I}{I_0}\right)$$

n is the intensity level in bels of a sound of intensity I referred to the threshold intensity I_0. The standard used for I_0 is the intensity of the quietest sound that the ear can detect; this is taken as 10^{-12} W/m^2. Thus an intensity of 10^{-8} W/m^2 can be termed an intensity level of 4 bels.

$$n = \lg\left(\frac{10^{-8}}{10^{-12}}\right)$$

The bel is rather a large unit and the decibel is more often used.

$$1 \text{ bel} = 10 \text{ decibels (dB)}$$

Because the ear is not equally sensitive to all frequencies a 50-dB sound at one frequency can be quite loud but at another it is barely audible. The **decibel scale** is not a loudness scale but an intensity scale. Loudness can be defined with reference to standards as it is possible to say when two sounds are of the same loudness. The standard used is the intensity level in decibels for a 1,000-Hz sound. Thus, if a sound at some frequency is equally as loud as a 30-dB sound at 1,000 Hz, its loudness is said to be 30 phons. Another

---

Experiment 9–3
Using two loudspeakers connected to separate signal generators or a pair of headphones with each earpiece connected to a separate generator, measure the current and the potential difference for each earpiece or loudspeaker. Set one generator at a standard level. Adjust the other generator to give a note which sounds twice as loud. Compare the power supplied to each loudspeaker or earpiece.

If possible set the standard at 1,000 Hz and 40 dB.

scale that is sometimes used for loudness is the **sone scale.** Over the range 20 to 120 phons the loudness in sones is given by

$$S = 2^{(P-40)/10}$$

where S is the loudness in sones and P the loudness in phons. One sone is defined as the loudness produced by a note of 1,000 Hz at 40 dB above a person's threshold. This gives an arithmetic scale for loudness as opposed to the logarithmic scale based on phons. If the sound is twice as loud as a 1-sone sound its loudness will be 2 sones, on the phon scale a change of 10 phons. (Experiment 9–3.)

9–2 Acoustics of buildings

In a similar way to water waves spreading out from a disturbance on a water surface, sound waves spread out from a vibrating object. In the case of sound the intensity, or amount of energy passing per second through unit area, varies inversely as the square of the distance between source and detector. The energy at a distance r from the source can be considered to have spread out over a sphere of surface area $4\pi r^2$, and hence the energy passing per second through unit area of the sphere is $1/4\pi r^2$ of the total energy spread over the surface. This is only true if the source of sound is in a region where no reflections can occur and the sound is received directly from the source. This condition is only really met out of doors away from any buildings or other reflectors. In a room reflections from many surfaces, walls, floor, and ceiling, etc., occur and the intensity at any point in a room is the summation of all these reflections and the direct sound. This generally gives a higher level of sound than would be obtained outdoors. Because all these reflections will not have the same path length the sound will persist over an interval of time. If there are reflections which produce a considerable path difference, two sounds may be heard, the direct sound and the reflected sound or echo.

(Experiments 9–4 and 9–5.)

In a room there are many reflecting surfaces and the sound at any point is due to a large number of reflections. The laws of reflection for sound waves are the same as those for any wave, e.g. light, and thus concave surfaces focus sound while convex surfaces cause it to diverge. Concave surfaces are generally undesirable in a room because the sound energy becomes concentrated in one particular region instead of being spread uniformly over the entire room. In the design of a room used for music or lectures, consideration must be given to the shape of the room in order to ensure that all sections of the audience receive an adequate level of

---

Experiment 9–4

Because of the similarity in behaviour between sound and water waves a study of the behaviour of water waves in an enclosure can lead to information concerning the behaviour of sound waves. With the aid of metal strips assemble an outline of a room in a ripple tank. A source of waves can be produced by allowing water to drip from a tube into the water at the appropriate point. Comment on the behaviour of waves in an enclosure.

Try the effect of changing the shape of the 'room'. Make the ceiling concave by using a curved metal strip.

Use convex surfaces for walls. If possible simulate an actual room. Comment on the results.

Experiment 9–5

Connect a microphone to an oscilloscope. Produce a short sharp sound, e.g. a toy cap exploding or two pieces of wood banged together. How does the sound at a particular point decay with time? Try different rooms.

sound. This can be achieved by the use of reflecting surfaces to direct sound to the otherwise low-level regions. Too great a path difference between the direct and the reflected sound must be avoided or the sound will be prolonged and it will be difficult to distinguish between successive notes of music or syllables of speech.

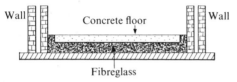

Wall   Concrete floor   Wall

Fibreglass

Fig. 9–2 Details of a floor mounting

The intensity of sound received at a particular point decays exponentially with time after the cessation of the sound. **Reverberation** is said to occur. The reverberation time is the time taken for the sound to fall in intensity by a factor of one million, that is 60 dB, after the cessation of the sound. The optimum reverberation time depends on the size of the room and the purpose for which it is to be used. For music reverberation times of the order of 1 or 2 s are optimum while for speech the times will be of the order of 0·5 to 1 s. These optimum times are those which allow successive notes or syllables to be distinguished but still allow a high sound level due

to reflection. Reverberation times can be adjusted by changing the amount of absorbing material present in a room. The more absorbers there are in a room, the shorter the reverberation time. As a guide the reverberation time is given by

$$t = \frac{V}{20A}$$

where t is the reverberation time, V the volume of the room in cubic feet, and A the total amount of absorption.

(Experiment 9–6.)

The amount of absorption can be calculated as the sum of the products of all areas and their absorption coefficients. The absorption coefficient of a material is $1 - r$, where r is the ratio of the sound energy reflected from a surface to that incident on it. Typical absorption coefficients are as follows.

Material	*Thickness (in)*	*Absorption coefficient at*		
		125	500	2,000 Hz
Brickwork	—	0·02	0·03	0·05
Acoustic plaster	½	0·15	0·35	0·60
Curtain	—	0·14	0·55	0·70
Expanded polystyrene	½	0·05	0·10	0·15
Carpet	⅓		0·15	0·60

Example 9–1. What is the reverberation time for a room with plaster walls and ceiling and a carpeted floor?

Experiment 9–6

Re-examine the results of Experiment 9–5 and determine the reverberation times for the rooms investigated. A photograph of the oscilloscope screen or the use of a pen recorder helps considerably. An approximation to the reverberation time can be obtained by determining the time taken for the sound to become inaudible. To obtain a more accurate figure a logarithmic graph is necessary to convert the exponential to a straight line.

The room's dimensions are 40 ft × 20 ft and it is 10 ft high. Take a frequency of 500 Hz.

$$t = \frac{V}{20A}$$

$$= \frac{40 \times 20 \times 10}{20 \times [0 \cdot 35 \times (40 \times 10 \times 2 + 20 \times 10 \times 2 + 40 \times 20) + 0 \cdot 15 \times 40 \times 20]}$$

$$= \frac{8,000}{820 \times 20}$$

$$= 0 \cdot 5 \text{ s.}$$

In addition to all these factors a room must be free of noise transmitted from outside. Noise can be transmitted from one room to another through the structure of the room (sound can travel through solids) or through the air. With structure-borne noise the sound energy is fed directly into the building structure as vibration. The structure's resulting vibration in other regions produces sound. One way of reducing this is to isolate the source of the sound energy from the room structure by putting the source on an elastic mounting. Care must be taken to ensure that no bolts or other fittings extend through the elastic mounting.

(Experiment 9–7.)

One method of eliminating structure-borne noise from concert halls is to use what is known as a floating floor. This is a floor in the form of a concrete raft laid on a Fibreglass quilt completely isolated from the rest of the building (Fig. 9–2).

Air-borne sound is sound carried through the air and partitions. The efficiency with which it is transmitted is reduced by solid barriers placed across the air path. As sound is carried by molecular motion of air, direct air links between rooms should be avoided. If the sound is being transmitted through a solid partition, increasing the density and hence the mass of the partition or using a partition of sound-absorbing material, e.g. expanded polystyrene, will reduce the effect.

9–3 Microphones and loudspeakers

A microphone is a device for converting sound signals to electrical signals. There are three main types, **electromagnetic, electrostatic,** and **piezoelectric.**

(Experiment 9–8.) The **electromagnetic** or **moving-coil microphone** consists of a small coil attached to a flexible diaphragm (Fig. 9–3) and placed between the poles of a permanent magnet. When sound is incident on the diaphragm, the pressure fluctuations cause the diaphragm to move and hence move the coil in the magnetic field. This induces an e.m.f. in the coil which

---

Experiment 9–7
Mount a motor directly on a bench or on the floor. Listen to the noise. Now place it in the same position but on a pad of foam rubber. What change occurs in the noise?

Experiment 9–8
Connect a wire across the terminals of a micro-ammeter (a spot galvanometer is suitable). Now move the wire up and down between the poles of a strong magnet. What happens?

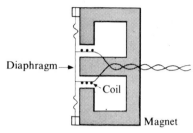
Fig. 9–3 Moving-coil microphone

varies with time in a similar manner to the pressure variations in the sound wave.

Another microphone depending on the same principle is the **ribbon microphone.** This consists of a strip of thin corrugated aluminium ribbon suspended between the poles of a permanent magnet (Fig. 9–4). The sound wave causes the ribbon to vibrate and hence produces

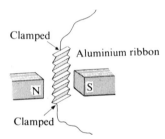
Fig. 9–4 Ribbon microphone

a voltage variation between its ends. This microphone is directional in that its maximum sensitivity is for a sound wave incident at right angles to its plane and almost zero for a sound wave incident along the plane of the ribbon.

The capacitance of a parellel-plate capacitor depends on the separation of the plates. Thus, if one plate is rigid and the other is flexible, a sound wave incident on

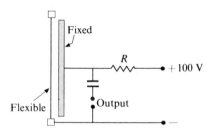
Fig. 9–5 Capacitor microphone

the flexible plate will produce changes in capacitance. This is the principle of the **capacitor microphone** (Fig. 9–5). A voltage of about 100 V is applied across the microphone and a high resistance in series with it. Changes in the capacitance cause potential-difference fluctuations between the terminals.

(Experiment 9–9.) The **carbon microphone** consists of a layer of loosely packed carbon granules between

---

Experiment 9–9

Connect a lead to a metal plate and place it at the bottom of a beaker. Fill the beaker with carbon granules. Place another metal plate on top of the carbon and attach a lead to it. Connect the leads from both plates to an Avometer switched to the resistance range or connect a milliammeter in series with the plates and a d.c. supply. What happens to the resistance of the carbon layer when pressure is applied to the top plate?

two plate electrodes (Fig. 9–6). Movement of the top electrode due to an incident sound wave causes resistance and hence current fluctuations. This is the type generally used in telephones where the quality of the sound is not so important. However, it is becoming superseded by the moving-coil microphone.

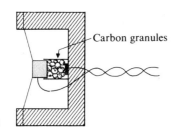

Fig. 9–6
Carbon microphone

The **crystal microphone** consists of a thin strip of suitable crystal which is subjected to mechanical stress by the movement of a metal diaphragm (Fig. 9–7). When certain crystals are mechanically stressed a potential difference is produced between opposite faces of the crystal. This occurs because the crystal changes shape under the action of the mechanical stress, and atoms of the crystals are thereby caused to move. Now, if these atoms are not neutral atoms but possess electric charge (they are then called **ions**), their movement causes the

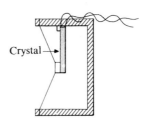

Fig. 9–7 Piezoelectric microphone

crystal surface to become charged and hence the appearance of a potential difference. This potential difference corresponds to the movements of the diaphragm caused by impinging sound waves. Typical piezoelectric crystals are quartz, Rochelle salt, and ammonium dihydrogen phosphate (ADP).
(Experiment 9–10.)

Basically **loudspeakers** are moving-coil microphones working in reverse. The moving-coil loudspeaker consists of a coil which can move in the space between the poles of a permanent magnet (Fig. 9–8). Instead of a metal diaphragm a large paper diaphragm (a cone) is used to move a large volume of air.

(Experiment 9–11.) When a current passes through the wire, the wire moves owing to the interaction of the magnetic field and the current. Thus, if a fluctuating

---

Experiment 9–10
Connect two electrodes to opposite faces of a Rochelle salt crystal. Aluminium foil can be used and attached to the crystal by means of Vaseline. Take the leads from the crystal to an oscilloscope or valve voltmeter. Gently tap the crystal. What happens? If the crystal is thin enough, try bending it.
 Suitable crystals can easily be grown. (See *Crystals and Crystal Growing* by Alan Holden and Phylis Singer, Heinemann Educational Books Ltd.)

Experiment 9–11
Connect two pieces of stiff bare copper wire to the terminals of a high-current low-voltage source (see Experiment 5–6). Complete the circuit by placing another piece of bare wire across the other two. The wires should be set horizontally. Pass the current through the wires. What happens? Bring up a strong magnet so that the direction of the magnetic field is at right angles to the plane of the wires. What happens?

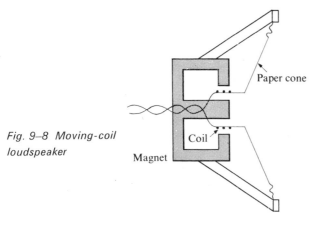

Fig. 9–8 Moving-coil loudspeaker

Paper cone

Coil

Magnet

current is fed to the coil of a loudspeaker, the cone of the speaker will move in accordance with the current changes.

9–4 Sound reproduction

There are basically three ways of 'storing' sound for later reproduction: (a) by cutting grooves on a disk, (b) by recording magnetic signals on tape, and (c) by recording optical signals on film.

(Experiment 9–12.) To produce a **record** a disk of a soft plastic is rotated on a turn-table at the required speed and a sharp cutting tool allowed to mark the surface. The cutting tool is made to oscillate, in a similar manner to the way in which the cone of a loudspeaker oscillates. The tool thus cuts a wavy pattern on the disk. The sound can be reproduced from the record by the reverse process, that is a stylus following the grooves and causing the coil to move in a magnetic field. The resulting electrical signal is amplified before being fed to a loudspeaker. An alternative form of pick-up head employs the piezoelectric effect and a suitable crystal is stressed by the movement of the stylus in the record groove and a potential difference generated.

(Experiment 9–13.) **Recording tape** consists of a layer of magnetic material on a plastic backing. As the tape moves past the recording head, an electromagnet, it becomes magnetized along the length. The magnetic material on the tape behaves in a similar manner to iron filings, and thus a magnetic record of the sound is produced (Fig. 9–9). The recording tape can be played back by the reverse process, that is the magnetic field from the moving tape inducing an e.m.f. in a pick-up coil connected to an amplifier and loudspeaker.

(Experiment 9–14.) The **sound track** on a film consists of a blackened region with either a variable-width transparent piece or a variable degree of blackness. In either case the light passing through the film varies in intensity in relation to the sound. These variations in

---

Experiment 9–12
Examine the grooves on a record with the aid of a magnifying glass or microscope.

Experiment 9–13
Connect an electromagnet, that is a coil of wire wrapped round a core of soft iron, to a switch and a d.c. supply. Now move over the top of the vertical coil a sheet of paper on which iron filings have been sprinkled. Switch the magnet on and off. What happens to the iron filings?

Experiment 9–14
(a) Examine a piece of cinema film. The sound track is down one edge.
(b) Direct a narrow beam of light onto a barrier-layer photocell. Connect the cell directly to the terminals of a microammeter. What happens to the current when the light is interrupted? Put the sound track in the light path and slowly move it through the light beam.

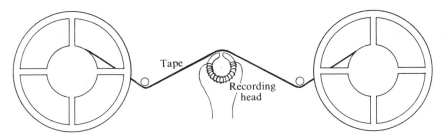

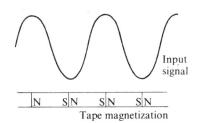

Input signal

Tape magnetization

Fig. 9–9 Principle of tape recording

light are converted by means of a photocell into an electrical signal.

Summary

Sound is a wave motion and is produced by vibrations of an object. The amount of energy passing through unit area in one second is known as the **sound intensity** and measured in watts per square metre (W/m²). A scale of intensity levels is given by

$$n = \lg\left(\frac{I}{I_0}\right)$$

where n is the intensity level in bels of a sound of intensity I. I_0 is 10^{-12} W/m². Ten decibels are equal to one bel. **Loudness** is defined in terms of phons and is the value in decibels of an equally loud 1,000-Hz sound.

In the design of buildings care must be taken to ensure no undue focusing of the sound but a reasonably even distribution throughout a room. Because of reflections the sound heard at any point in a room takes a finite time to decay and a measure of this is the **reverberation time.** This is the time taken for a sound to decay by 60 dB.

$$\text{reverberation time} = \frac{V}{20A}$$

where A is the total absorption and V the volume of the room. Another factor that must be taken into consideration for lecture rooms or concert halls is the elimination of noise from external sources.

There are basically three types of microphones, **electromagnetic, electrostatic,** and **piezoelectric.** The electromagnetic microphone makes use of the fact that a coil moving in a magnetic field generates an e.m.f.; the electrostatic microphone uses the variation in capacitance and hence in potential difference produced by varying the plate separation of a parallel-plate capacitor, and the piezoelectric microphone the e.m.f. produced across a crystal by mechanical stress. **Loudspeakers** are basically moving-coil microphones working in reverse.

Sound can be 'stored' for later reproduction by cutting grooves on a disk, by magnetizing a tape, and by imprinting varying patterns of light and shade on a film.

Problems

9–1 What is the intensity level in decibels if the sound intensity is 10^{-6} W/m²?

9–2 By what factor does the sound intensity change if the sound intensity level changes from 10 to 20 dB?

9–3 How does a 20-phon sound differ from a 20-dB sound?

9–4 Design a lecture room, about 40 ft × 20 ft and 10 ft high. Consider the problem of reverberation and an adequately uniform distribution of sound throughout the room.

9–5 How could the noise generated by a motor mounted on the floor of a laboratory be reduced?

*9–6   Explain how a record player works.

*9–7   Why does the size of a loudspeaker cabinet affect the sound produced?

9–8   How is sound produced?

Part 2

10 Light

10–1 Colour

White light when dispersed by a prism produces a continuous range of colours ranging from red, through orange, yellow, green, blue, to violet. These colours when combined constitute white light.

(Experiment 10–1.) A red surface when viewed in white light normally appears red because the molecules comprising that surface absorb all the colours of the light incident on it other than the red which is reflected to the eye of the observer. A white surface reflects all colours while a black surface reflects none.

(Experiment 10–2.) Colour filters absorb some colours and transmit others. Thus a red filter absorbs all the spectrum apart from the red. A green filter absorbs the blue and the red parts of the spectrum and allows the green to pass through.

(Experiment 10–3.) A white patch is produced on a screen where red, green, and blue lights overlap. Effectively the spectrum may be considered to consist of three sections—the red, the green middle region, and the blue. When the three coloured lights are added they give all the colours in the spectrum and hence the colour is white. These three colours are known as the **primary colours.** The colours produced by the addition of two primary colours, e.g. yellow from red and green, magenta from blue and red, turquoise from blue and green, are known as **secondary colours.**

Combining coloured lights to give various colours is an additive process. The colours of paints are due to a subtractive process in which the paint absorbs all but one region of the spectrum. For example, a blue paint reflects blue light but absorbs other colours. Paints are, however, not pure colour absorbers, and so the blue paint will reflect a little of the colour next to it in the spectrum, that is green. A yellow paint reflects the orange, yellow, and green. Thus, when blue and yellow paints are mixed, the result is green as this is the only colour which both will reflect. If blue and yellow lights are mixed, the result is white.

10–2 Photography

In the **camera** an image of the object being photographed is produced on a light-sensitive emulsion. This records a latent image which can be revealed by chemically processing the emulsion. The result is known

---

Experiment 10–1

In a darkened room direct white light onto red, green, and blue surfaces in turn. What is the appearance of the surfaces? Now direct a beam of red light onto each surface and observe the colours. Repeat this with green and with blue light.

Try the experiment with white and black surfaces. Why do surfaces assume the colour we associate with them?

Experiment 10–2

Place a red filter between a white lamp and your eye or a white screen. What is the colour of the light transmitted by the filter? Repeat the experiment with a green filter and then a blue one.

Experiment 10–3

Allow light from three lamps to pass separately through three filters, red, green, and blue. Direct the three coloured beams of light onto a white surface so that the three patches partly overlap. What happens when all three overlap? What happens when red and blue, green and red, and blue and green overlap?

as a **negative.** Essentially this is the photographic process.

(Experiment 10–4.) Essentially a camera consists of a lens and the emulsion. The distance from the lens to the emulsion can be varied, as also can the size of the aperture through which the light passes into the camera (Fig. 10–1). In addition the interval of time during

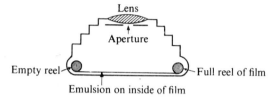

Fig. 10–1 Basic features of a camera

which the light can enter the camera is controlled by a shutter. The number obtained by dividing the focal length of the lens f by the effective diameter of the aperture through which the light enters the camera is known as the **'f number'.** Thus, if a lens has a focal length of 10 cm and an aperture of 2·5 cm, then the f number is $f/4$. The f number can be varied by altering the size of the aperture. A typical series of f numbers would be $f/2$, $f/2·8$, $f/4$, $f/5·6$, $f/8$, $f/11$, $f/16$, $f/32$. The values of the f numbers are chosen so that, in going from one number to the next higher number, the light passing through the lens is reduced by half. The amount of light passed by a lens is inversely proportional to the square of the f number. The greater the f number, that is the smaller the aperture, the greater is the depth of the field of view which is in focus, e.g. $f/16$ may give all the objects between 3 m and infinity in focus, whereas $f/4$ would focus only on objects between 3 and 4 m distant.

A photographic film or plate consists of a layer of silver bromide on a base of glass or a cellulose ester. The emulsion, silver bromide in the form of fine crystals or grains, is embedded in gelatin. When light falls on the emulsion, an image is formed. On development the image in converted to a visible image of silver. The action of the light is to make a grain which has received light capable of being attacked by the developer. The size of the grains in the emulsion determines the coarseness of the negative and also the speed of the film, that is the sensitivity to light. The coarser the grain, the more sensitive is the emulsion to light. The normal development procedure consists of a number of steps.

(a) Immersion of the film in the developer for a definite interval of time. The length of time depends on the emulsion used, the developer, and the temperature. Details will be found with the developer used.

Experiment 10–4

Make a box with one end covered by a sheet of tracing paper and the opposite end covered by a sheet of black paper. Make a small hole in the black paper and observe the tracing-paper back when the pinhole is pointed at a bright object such as a lamp in a darkened room. Make more holes in the paper. What is the result on the image seen on the tracing paper?

Put a convex lens in contact with the hole in the black paper. What is the result now?

With a new piece of black paper study the effect of increasing the size of a single central hole, both with and without the lens. Try objects at different distances.

(b) Washing. This is a quick rinse in water to remove the surplus developer and slow the progress of the development.

(c) The emulsion is then immersed in a fixing bath. This stops development and removes the residual silver bromide. Up to this point the processing would generally have been carried out in complete darkness. This is most conveniently done in what is known as a developing tank. Liquids can be admitted and withdrawn from the tank without daylight entering. The only part of the procedure which generally requires a darkroom is the loading of the film into the tank. After an appropriate time in the fixing bath the film is given a final wash. This is to remove all the soluble salts left in the emulsion and will generally last about 30 min.

Positive prints from the dried negative can be obtained by passing light through the negative onto a piece of light-sensitive paper. If enlargements are required, an image of the negative is projected onto the sensitive paper. The development procedure is the same as for the film; the paper can, however, be handled in an appropriately coloured light, generally yellow, normally called a safe-light.

(Experiments 10–5 and 10–6.)

Experiment 10–5

Take a photograph and develop the film. The exposure should be determined by the use of an exposure meter. If no meter is available, use the guide given by the film manufacturers and photograph the same scene with a number of different exposures.

Experiment 10–6

By the use of the appropriate safe-light make prints of a negative.

Summary

White light can be considered to be composed of three **primary** colours—red, green, and blue. Combining coloured lights to give other colours is an additive process, whereas mixing paints has a subtractive effect on the colours. Paints absorb certain colours and their resultant appearance is due to the remaining reflected light.

In the **camera** an image of an object is produced on a light-sensitive emulsion by means of a lens. The image is made permanent in the emulsion at a later stage in the development procedure. The amount of light falling on the film is governed by the shutter speed, the f number, and the brightness of the object being photographed. The f **number** gives a measure of the size of the camera aperture, and the bigger the number the smaller the aperture.

Problems

*10–1 How does the human eye see colours? How does colour television work?

10–2 What is the effect on the amount of light falling on a film if the aperture is changed from $f/4$ to $f/8$? What else changes?

10–3 What is the effect on the photograph produced if the speed of the film is increased?

10–4 Explain how you would photograph the following: (a) a large machine in the laboratory, and (b) the waves in a ripple tank. Full details of the focusing and choosing of aperture and shutter speed should be given.

11 Light Waves

11-1 Interference and diffraction

Two of the fundamental characteristics of water waves are **interference** and **diffraction.** With interference two waves meeting at a point can give a maximum or minimum displacement or some value in between. Diffraction is the spreading out or 'bending' of a wave round corners. Particles would not be expected to give either of these properties—when two particles meet the result is always two particles; also a beam of particles does not normally bend round corners.

Interference and diffraction of light are not obvious phenomena—light appears to travel in straight lines and does not bend round corners; two beams of light do not meet and give darkness. One point does, however, emerge from a study of water waves (see Chapter 2)—diffraction and interference effects are only easily detected when we are working with objects whose dimensions are of the same order as the wavelength. For example, diffraction through a slit of width 10λ is much less pronounced than that through a slit of width 1λ. Perhaps the wavelength of light is very small. Again with interference between waves from two sources the maxima and minima are quite far apart when the two

sources are only a matter of a few wavelengths apart but very close together when the separation is more than about 10λ (see Chapter 2). Thus, if the wavelength of light is small, we would expect interference to be most noticeable when two sources are separated by a distance comparable with the wavelength, that is a very small distance.

(Experiments 11–1 and 11–2.) These experiments show that light behaves as a wave motion with a wavelength of the order of 5×10^{-7} m. Light will 'bend' round corners and two light waves when added together can give darkness. The results are comparable with those produced by water waves.

When light 'bends' into the shadow region, uniform illumination is not produced owing to interference between the diffracted waves. In some parts of the shadow the waves cancel each other and give darkness, while in other parts of the shadow they add up to give brightness. Thus in the case of the shadow of the ball bearing a bright spot is produced in the centre of the shadow owing to constructive interference between the waves bending round the edges; each wave travels the same distance and thus they will arrive in phase.

Experiment 11-1

The room should be darkened in order that shadows cast by the light should be clearly visible. A very bright source of light should be used; a tungsten iodide lamp is ideal. The shadows cast by various objects should be viewed through a translucent screen about a metre away (Fig. 11–1). Try as objects a straight edge such as a razor blade, a ball bearing of diameter about $\frac{1}{4}$ in, small holes in a metal sheet, a hair, etc. In all cases focus your eyes on the image on the screen and observe both inside and outside the shadow. Is light behaving as a wave motion?

Fig. 11–1

Experiment 11-2

Draw two fine slits about a millimetre apart to produce two very closely spaced identical sources of light. A con-

In the case of interference between waves originating from the two slits a system of parallel regions of constructive and destructive interference occurs (Fig. 11–2). Each bright fringe represents a position where two waves arrive in phase, that is crest meets crest and trough meets trough, while each dark fringe is due to waves being out of phase, that is crest meets trough. An important point regarding interference of light is that we cannot obtain interference effects between light waves originating from two separate sources. This is because the light is not coherent, that is the light from each source comes from a large number of atoms, each of which sends out a short-duration pulse of light at random intervals. Although we may obtain interference between waves from two separate sources, it will only last for a very short time and will be impossible to see. It is as though in the experiment with the ripple tank we kept stopping and starting each vibrator in a random manner—we would never detect interference. By the use of two slits, however, we divide the light from one source into two parts and thus both slit sources emit in the same manner and it is possible to obtain interference.

Consider light originating from two slits S_1 and S_2 and arriving at a point P (Fig. 11–3). The light will be in phase if crest meets crest and trough meets trough. The two waves, however, travel different distances; the

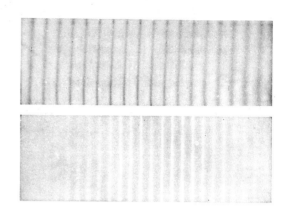

Fig. 11–2 Interference patterns of red light (top) and blue-violet light

wave from S_2 travels an extra distance S_2T. The waves will be in phase if this distance is a whole number of wavelengths. Now

$$\frac{S_2T}{S_1S_2} = \frac{PQ}{PS_2}$$

This assumes that the distance S_2Q is very large compared with S_1S_2; the light from the two slits is considered to be virtually parallel. If PQ is the fringe

venient method of doing this is to coat a microscope slide with Aquadag and then to cut through this layer with a fine needle. In order to produce slits very close together and at a known separation a piece of apparatus can be used which enables you to draw one slit by running the needle along a metal straight edge and then the slide is moved relative to that edge by a known amount by means of a screw (Fig. 11–4).

Place the double slit about a metre away from a straight-filament lamp; the slits should be parallel to the filament. Place the translucent screen a few metres away from the slits and observe the light on the screen. What can be seen? How does the result compare with that produced by water waves either passing through two slits or produced by two vibrators? Obtain a rough estimate of the wavelength of light. What happens if different colour filters are placed between the lamp and the slits?

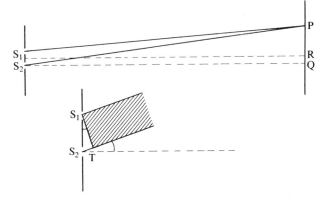

Fig. 11–3 Interference produced by two illuminated slits

separation, that is the distance between the centre of one bright fringe and the centre of the next bright fringe, then S_2T represents a path difference of one wavelength. Thus

$$\frac{\lambda}{S_1S_2} = \frac{\text{fringe separation}}{\text{screen to slit distance}}$$

S_2Q is almost equal to PS_2.

Example 11–1. Calculate the wavelength of light which with a double-slit arrangement produces fringes 2·4 mm apart on a screen 2 m away from slits of separation 0·5 mm.

Using the above formula

$$\frac{\lambda}{0·5 \times 10^{-3}} = \frac{2·4 \times 10^{-3}}{2}$$

All the distances are expressed in metres. Hence

$$\lambda = 6 \times 10^{-7} \text{ m}$$

One obvious deduction from the formula is that the fringe separation depends on wavelength. Thus with white light a series of overlapping fringes is produced with only the centre of the pattern, R, constructively interfering for all wavelengths; the centre fringe is white since all the waves have travelled the same distance.

11–2 Thin films

Another method of obtaining interference with light waves is to divide the wave into two parts by reflection at a surface. Consider a thin soap film. This shows a series of colours due to interference of white light. The

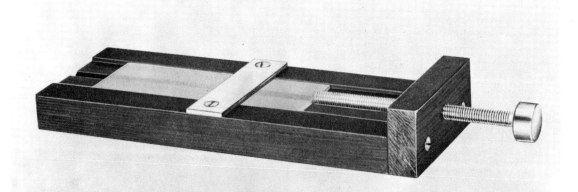

Fig. 11–4
Double slits
apparatus

interference is produced between the waves reflected from the front and back surfaces of the film. (Experiment 11–3.)

With the double-slit experiment interference is produced because of a path difference between the two beams of light; with a thin film the path difference occurs by virtue of the light reflected from the top of the film interfering with the light from the bottom of the film—again a path difference. Consider a film of thickness d and refractive index μ, with light incident normally on the film (Fig. 11–5). There is a path difference of

Fig. 11–5 Interference produced by thin film

$2d$; however, because this path difference is in a medium of refractive index μ, we cannot use the wavelength in air to find out whether this difference is a whole number of wavelengths.

$$\mu = \frac{\text{velocity of light in air}}{\text{velocity of light in medium}}$$

But the velocity is the frequency multiplied by the wavelength and, as the frequency is constant,

$$\mu = \frac{\text{wavelength in air}}{\text{wavelength in medium}}$$

Thus, if the path difference is the wavelength in the medium,

$$2d = \text{wavelength in medium}$$
$$2d = \frac{\text{wavelength in air}}{\mu}$$

We might expect that this would give constructive interference as d tends to zero, but this is not so. When the soap film was very thin in the experiment, that is d very close to zero, it appeared black, that is the light reflected from the front surface gave destructive interference with the light reflected from the back. But the two surfaces were very close together—thus apparently, with virtually zero path difference, destructive interference occurs. The only possible explanation is that on reflection at one of the surfaces a half-wavelength path difference must have appeared. Thus for constructive

---

Experiment 11–3

Make a wire loop and arrange it vertically. Dip the frame into a beaker of soap solution or liquid detergent (Stergene diluted 1 in 100). Observe the film by reflected light. Explain what is seen. What happens as the film becomes thinner? What is the colour of the light reflected from a very thin film? Is this last colour to be expected?

interference the smallest path difference is when

2d + ½(wavelength in the medium)
                    = wavelength in the medium

$$2d + \tfrac{1}{2}\lambda_m = \lambda_m$$
$$2d = \tfrac{1}{2}\lambda_m$$

but

$$\lambda_m = \lambda_a/\mu$$
$$2d = \tfrac{1}{2}\,\lambda_m/\mu$$

therefore

$$2\mu d = \tfrac{1}{2}\,\lambda_a$$

where λ_a is the wavelength in air, and λ_m is the wavelength in the medium. Further constructive interference will occur when the film is thicker, that is when

$$2\mu d = (n + \tfrac{1}{2})\lambda_a$$

where n is an integer. Destructive interference will occur when

$$2\mu d = n\lambda_a$$

Example 11–2. Calculate the minimum thickness of a soap film which will appear bright when viewed by reflected light of wavelength $5 \cdot 2 \times 10^{-8}$ m. The refractive index of the soap solution may be taken as $1 \cdot 3$.

$$2\mu d = \tfrac{1}{2}\lambda$$

Hence

$$d = \frac{5{,}200 \times 10^{-10}}{2 \times 2 \times 1 \cdot 3}$$
$$d = 10^{-7}\ \text{m}$$

If the film thickness varies, that is d varies, then a series of fringes will be seen which are effectively 'contour lines', marking out the positions of equal film thickness. Thus for a wedge the fringes are parallel to its edge. A difference in film thickness of $2d$ occurs between successive fringes.

(Experiment 11–4.)

Circular fringes are produced when a convex lens of large radius of curvature is placed in contact with a piece of plane glass. These fringes are known as **Newton's rings.** If r is the radius of such a fringe and R is the radius of the convex lens, then (Fig. 11–7)

$$r^2 = (2R - d)d$$

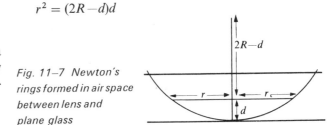

Fig. 11–7 Newton's rings formed in air space between lens and plane glass

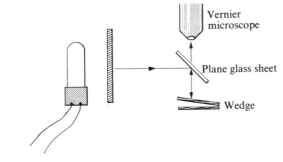

Fig. 11–6

Vernier microscope

Plane glass sheet

Wedge

Experiment 11–4

Clean two pieces of plate glass and form an air wedge between the plates by inserting a thin piece of paper along one edge. View the light reflected from the surface of the wedge when illuminated by an extended source of monochromatic light; a sodium lamp or Bunsen flame containing sodium placed behind a piece of tracing paper is suitable (Fig. 11–6). A convenient method of observing the fringes is to reflect the light onto the wedge by means of a piece of glass placed at 45° to the incident light from both the source and the wedge.

Describe what is seen. Determine the thickness of

d is the thickness of the air film giving rise to the fringes. If $R \geqslant d$ we can neglect d^2; hence

$$r^2 = 2Rd$$

Now for a dark fringe

$$2\mu d = n\lambda$$

Hence

$$r^2 = \frac{Rn\lambda}{\mu}$$

It is often difficult to determine n owing to the closeness of the rings near the centre. Hence a common procedure is to ignore the smallest rings and begin counting with a ring which is easy to see. Thus this ring is taken as number 1, the next as number 2, and so on. The actual number n of the ring is given by

$$n = n_1 + C$$

where n_1 is the number of the ring when the inner rings are ignored and C is a constant, the number of rings ignored. Substituting this in the equation gives (note that in air $\mu = 1$)

$$r^2 = R\lambda(n_1 + C)$$
$$r^2 = R\lambda n_1 + R\lambda C$$

Plotting r^2 against n_1 gives a line whose slope is $R\lambda$, from which a value of the wavelength can be obtained. (Experiment 11–5.)

11–3 The diffraction grating

With two closely spaced slits interference fringes are produced—what happens if we have more than two slits? If the slits are very narrow, then in a plane normal to the slits each behaves as if it were a point source (examine this with water waves), and thus we would expect to get interference between the light coming from each source. If each wave is to start in phase, then we must have parallel light incident on and normal to our array of slits. An array of very fine slits is known as a **diffraction grating.**

(Experiments 11–6 and 11–7.)

When each slit in Experiment 11–7 acts as a source of waves an apparently confused pattern of interference occurs immediately after the slits, but at some distance away plane waves are found emerging in particular directions. These plane waves are actually the portions of the circular waves that are along the tangents linking the wave from each source (Fig. 11–8). It is along these directions that constructive interference occurs—the other parts of each circular wave interfere destructively. Consider the waves coming from a particular pair of

the paper by counting the number of fringes between the edge of the wedge and the paper. Near the paper this may be difficult and the number can be estimated by counting the number of fringes per centimetre along the wedge and measuring the distance from the paper to the edge.

Experiment 11–5
Place a lens of long focal length on a piece of plane glass and direct light onto it normal to the plane glass. An extended source of monochromatic light should be used.

Determine the wavelength of the light used.

Plane wave-front built up from portions of
spherical waves

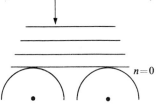 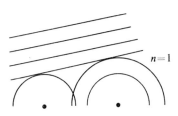 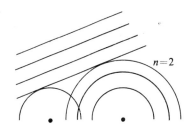

Fig. 11–8 The directions in which constructive interference occurs

slits at an angle θ (Fig. 11–9). If the path difference
between the waves from each slit is a whole number of
wavelengths, constructive interference can occur. Thus
for $RQ = n\lambda$, where n is an integer, constructive interference will occur. But

$$RQ = PR \sin \theta$$

Hence

$$PR \sin \theta = n\lambda$$

θ is the angle through which the waves have been
diffracted. PR is the separation of the slit centres. If
there are N slits per unit length of grating, then

$$PR = \frac{1}{N}$$

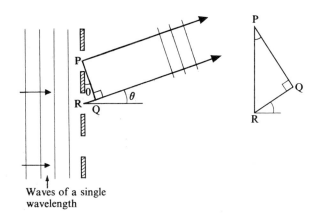

Waves of a single
wavelength

Fig. 11–9 Diffraction of light by a grating

Experiment 11–6
Light from a slit or a line-filament lamp should be made
parallel by placing a convex lens at a distance from the
slit equal to its focal length. Direct the light onto a
diffraction grating. Examine the light reaching a distant screen. Try putting a lens between the grating and
the screen to focus the light onto the screen. Try the
experiment with both fine and coarse gratings.

Experiment 11–7
Assemble a ripple tank with a large number of either
equally spaced vibrators or slits. Observe the resulting
interference between the water waves. Compare the
results with those of Experiment 11–6. The waves
should be examined close to the slits and at some
distance away.

Hence for constructive interference

$$\frac{1}{N} \sin \theta = n\lambda$$

Example 11–3. A diffraction grating is ruled with 5,000 rulings/cm. What will be the angle between the zero and first orders for a wavelength of 4×10^{-7} m?

n is known as the order number and thus the angle required in this problem is that between the first two fringes.

$$\frac{1}{N} \sin \theta = n\lambda$$

Hence

$$\sin \theta = 5{,}000 \times 100 \times 1 \times 4 \times 10^{-7}$$
$$\sin \theta = 0{\cdot}20$$
$$\theta = 11° 32'$$

(Experiment 11–8.)

11–4 Spectra

When the light from a glowing solid or liquid is examined using either a diffraction grating or a prism, a continuous spread of colour is observed, that is there is a continuous variation of wavelength. This is known as a **continuous spectrum.** If, however, a hot gas is examined,

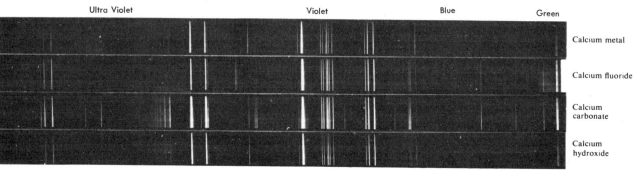

Fig. 11–10 The spectrum of calcium

Experiment 11–8
Determine the wavelength of light emitted from a sodium flame or lamp. This can be done using the arrangement described in Experiment 11–6. The angle θ can be determined from measurements of the distance from grating to screen and the distance between the zero-order and the various-order fringes on the screen. N is supplied by the grating manufacturer.

the spectrum is found to consist of a number of coloured lines, that is discrete wavelengths. This is known as a **line spectrum** (Fig. 11–10). The line spectrum is characteristic of the element producing it—no two elements have the same spectra. Thus, when sodium chloride is heated in a Bunsen flame, the spectrum produced is characteristic of sodium; the chlorine spectrum cannot be easily seen. These spectra from hot substances are called **emission spectra.**

(Experiment 11–9.)

If certain wavelengths in white light are absorbed when the light passes through a substance, the resulting spectrum is called an **absorption spectrum.** It consists of the spectrum due to the white light, with dark lines where the absorbed lines are missing.

(Experiment 11–10.) Spectroscopy, involving either line emission or absorption spectra, is widely used to identify the materials present in a mixture and to determine the amounts. With the emission spectra the identity of an element can be determined from a knowledge of the wavelengths emitted; the quantity can be determined from measurements of the intensity of the lines. With absorption spectra the regions absorbed give clues to the identity of the molecule concerned and the amount of absorption determines the concentration present.

11–5 Polarization of light

Two types of wave motion are possible—**transverse** and **longitudinal.** In the transverse wave the displacement of the wave is at right angles to the direction the wave is travelling, while in a longitudinal wave the displacement is along the direction the wave is travelling. With a longitudinal wave only one possible mode of displacement is possible; with a transverse wave there are many possible planes of displacement. Thus, if it were possible to have a device which was sensitive to just one plane of displacements, then this would distinguish between the two types of wave.

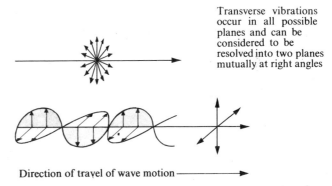

Transverse vibrations occur in all possible planes and can be considered to be resolved into two planes mutually at right angles

Direction of travel of wave motion ⟶

Fig. 11–11 Transverse waves

Experiment 11–9

Arrange the diffraction grating to produce a spectrum on a screen. Mark on the screen the position of the yellow line in the sodium spectrum. Now place various materials in a Bunsen flame and determine whether any sodium is present, e.g. a glass rod, some soil.

An alternative to assembling the diffraction grating is to use a commercial direct-vision spectroscope. This has an internal scale and either a grating or prisms to produce the dispersion.

Experiment 11–10

Use either the direct-vision spectroscope or the diffraction-grating assembly and view the spectrum produced by a hot solid, a filament in an electric light bulb. Now place between the lamp and the spectroscope a beaker or test-tube containing water. Is there any change in the spectrum? Add chlorophyll or potassium permanganate or a few drops of ink to the water—is there any change in the spectrum? Does the result depend on the concentration?

(Experiments 11–11, 11–12, and 11–13.) Normal white light is composed of (a) light of various wavelengths, and (b) transverse waves in many planes of vibration. It is possible to represent these different planes by two planes mutually at right angles. These are called **planes of polarization** (Fig. 11–11). When light passes through a single sheet of Polaroid, vibrations in one of these planes are absorbed; the vibrations that are left are said to be **plane polarized.** The eye is not normally sensitive to the orientation of the plane of polarization of light and so no difference is detected by the eye when the light is viewed through a single Polaroid sheet. If a second sheet of Polaroid is now placed between the first sheet and the eye, this will only pass light if the plane of polarization is in the right direction. As the light coming through the first Polaroid is now plane polarized, that is the vibration is in one plane only, the second Polaroid will pass this light when its axis is in line with the plane of polarization of the light, and not when its axis is at right angles to the plane of polarization. Thus, as the Polaroid is rotated, the light intensity will pass through alternate maxima and minima for each 90° of rotation (Fig. 11–12).

When unpolarized light is reflected from a piece of glass the light will be found to be partially plane polarized and at one particular angle of reflection completely

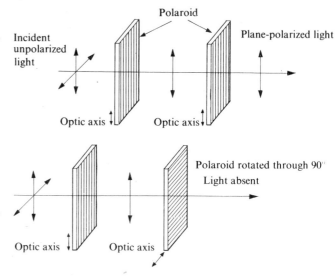

Fig. 11–12 The passage of light through a pair of Polaroid sheets

plane polarized (Fig. 11–13). This is the principle behind the use of Polaroid sunglasses. They reduce the glare from sunlight reflected from water or roads, etc., by removing the light that has been polarized by reflection.

Calcite produces what is known as **double refraction,**

Experiment 11–11

Take a piece of Polaroid and rotate it in front of the eye so that the light passing through the Polaroid is viewed. What happens? Is light a longitudinal vibration, a mixture of transverse waves, or just a single transverse wave?

Place a second piece of Polaroid between the first piece and the light. Rotate one of the Polaroids. What conclusion can you draw from your observations?

Experiment 11–12

Observe light reflected from a piece of glass or from the laboratory bench through a piece of Polaroid. Rotate the Polaroid. What happens? What conclusion can you draw?

Experiment 11–13

Place a calcite crystal over a point marked with a pencil on a piece of paper. Rotate the crystal. What is seen? Place a piece of Polaroid over the crystal and rotate the Polaroid. What happens? Explain your observations.

that is two refracted beams are produced for one single incident beam. Each of the refracted beams is plane polarized, at right angles to the other. By suitably cutting and joining two pieces of calcite crystal it is possible to produce a device for giving plane-polarized light. This is known as a **Nicol prism.**

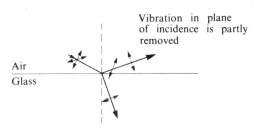

Fig. 11–13 Polarization by reflection

11–6 Interference of polarized light

(Experiment 11–14.) A transparent material which is normally colourless may, when put between a pair of Polaroids, appear coloured. This occurs when the light splits up into two plane-polarized components on refraction through the material. Now refractive index is the ratio of the speed of light in air to that in the material, so we conclude that the two refracted components have different speeds in the material. Thus in passing through the material one ray will get ahead of the other, by an amount depending on the difference in refractive indices and the thickness of the material. The second Polaroid combines these rays and thus interference can occur between the two waves. When the path difference produces, for example, destructive interference for blue light, then the result will be white light minus blue light. This appears as an orange colour. There is thus a sequence of interference colours as the thickness of the sample increases and successive wavelengths are removed from the incident light. The first-order interference colours follow the sequence: white minus violet giving yellow; white minus blue giving orange; white minus green giving red; white minus yellow giving violet; white minus orange giving blue; white minus red giving green. For thicknesses greater than this the colours go through another sequence when two colours are extracted each time.

One application of this effect is the **polarizing microscope.** This is similar to a normal microscope but has two Polaroids, one below the microscope stage, where the specimens are placed, and one in the barrel of the microscope. These transform what was a view of a uniformly coloured substance into a multi-coloured image and so enable differences in structure to be more easily seen.

Experiment 11–14

Place a sheet of Cellophane or a strip of Sellotape between two pieces of Polaroid film. What is the effect of this on incident light? Rotate the Polaroid. Try more than one later of Cellophane or Sellotape. Try the effect of stressing the Cellophane or Sellotape. Comment on your observations. In all these cases use a white-light source.

(Experiment 11–15.)

When some materials are stressed they become **birefringent,** that is they become doubly refracting and have two refractive indices. The amount of birefringence depends on the strain in the material. Thus when a specimen between crossed Polaroids is stressed the amount and directions of the strain axes can be determined from the colours and their positions. This effect enables mechanical stresses to be studied, the subject being called **photoelasticity.**

(Experiment 11–16.)

Summary

Light behaves as a wave motion with wavelengths of the order of 5×10^{-7} m. Because of its wave nature light shows **interference** and **diffraction** effects. Since light is emitted randomly from the atoms in a source, interference between two separate sources is impossible to observe. Interference is seen only between two sources produced by a subdivision of one, e.g. by a double slit or by two reflections from a thin film. With reflections a phase change equivalent to a displacement in wavelength of $\frac{1}{2}\lambda$ occurs when light is reflected at a less dense to a more dense boundary (a high- to low-velocity

boundary). For destructive interference with light incident normally on a thin film of thickness d

$$2\mu d = n\lambda$$

With a convex lens on a plane glass plate the air film gives circular fringes known as **Newton's rings.** The fringes appear as contour lines with a height difference of $\lambda/2\mu$ between successive fringes.

A number of closely spaced slits constitute a **diffraction grating** and can be used to determine wavelengths. For a grating with N rulings per unit length

$$\frac{1}{N}\sin \theta = n\lambda$$

θ is the angle through which the light has been diffracted.

Incandescent solids and liquids emit **continuous spectra** which are the same for all materials; **line emission spectra** are obtained from hot gases and are characteristic of the elements producing them. When white light is passed through a material such as gas, certain wavelengths are absorbed—the resulting spectrum is known as a **line absorption spectrum.**

Light is a transverse wave motion and thus can be polarized, that is the mode of vibration can be restricted to a single plane. This can be done by **selective absorption,** by **double refraction,** or by **reflection.** When polar-

---

Experiment 11–15
Convert a simple microscope into a polarizing microscope. One Polaroid can be placed beneath the stage and one above the eyepiece. Examine a thin layer of ice. What is seen as one Polaroid is rotated?

Experiment 11–16
Flex a piece of celluloid between crossed Polaroids. Comment on the pattern of colours. Try flexing the celluloid.

Observe a piece of glass between crossed Polaroids. Then stress the glass by means of a G clamp and observe the resulting colours. Try squeezing a Perspex ring with a G clamp.

ized light is passed through certain materials, double refraction occurs and the two rays travel at different speeds through the material. When components of the two rays are combined by a Polaroid sheet, interference can occur. With white light this results in certain colours being missing. The difference in refractive indices can be produced in certain materials by subjecting them to mechanical stress; these materials are called **photoelastic materials.**

Problems

11–1 Two coherent sources separated by a distance of 1 mm radiate light of wavelength 5×10^{-7} m. What is the fringe separation on a screen 1 m in front of the sources?

11–2 Explain how reflection from a surface such as a glass lens can be reduced by coating the surface with a thickness of a quarter of a wavelength of suitable material. Carefully explain the reason for the particular thickness and the choice of the material.

*11–3 In geometric optics light is considered to travel in straight lines. In view of diffraction this would seem wrong. Is it? Explain your answer.

11–4 A very thin soap film appears black. Why?

11–5 A plane transmission diffraction grating has 4,000 lines/cm. At what angles will be the various orders for a wavelength of $5 \cdot 89 \times 10^{-7}$ m when it is illuminated at normal incidence with parallel light?

11–6 Two glass plates in contact along one edge are separated at the opposite edge by a piece of paper. When light of wavelength $5 \cdot 89 \times 10^{-7}$ m is incident along the normal to the wedge, 40 fringes are counted across the width of the plates. What is the thickness of the paper?

11–7 What is the minimum size an object can be if it is to be seen with light of wavelength 5×10^{-7} m?

*11–8 Why is interference of light not a phenomenon observed whenever two lights are switched on in a room or light is incident on a window?

11–9 Explain how Newton's rings are produced.

11–10 Write a note explaining the basis of photoelasticity.

11–11 How can spectroscopy be used to identify elements?

12 Electromagnetic Waves

12–1 Generation of electromagnetic waves

When a magnet moves past a wire, an e.m.f. is induced in the wire and an electric current will flow; this effect is called **electromagnetic induction.** The electric field set up in the wire causes a current to flow, which in turn produces another magnetic field. Thus a varying magnetic field generates an electric field, and conversely a varying electric field (from the flow of charge in the wire) generates a magnetic field. However, what would happen if the wire in which the current flows were removed? Would a varying magnetic field generate an electric field and vice versa? Surprising as it may seem this is indeed the case and, when the variation of the fields with time is great enough, such a system of electric and magnetic fields will propagate or travel through space. In a vacuum their speed will be 3×10^8 m/s, which is equal to the speed of light in a vacuum. Such a propagating system of fields is called an **electromagnetic wave.** These varying fields are only generated when electric charge undergoes acceleration. In practice a radio wave is generated by high-frequency currents flowing in a conductor—the **aerial.** In the case of radiated light the charges would be the electrons within the atom—when light is emitted by an atom, the charges are presumably accelerating. If no emission occurs, the electrons cannot be accelerating.

(Experiment 12–1.)

12–2 Radio waves

One method of accelerating charges would be to apply a very-high-frequency e.m.f. to a wire. Such oscillations can be produced by the discharge of a capacitor through an inductor. The oscillations are, however, damped and gradually die away.

(Experiment 12–2.) When a capacitor discharges through an inductor, damped oscillations are produced, the frequency f being given by

$$f = \frac{1}{2\pi(LC)^{1/2}}$$

L being the inductance in henries and C the capacitance in farads.

If the oscillations are to be maintained, then energy at the right frequency must be fed into the circuit. This can be done by inserting the oscillatory circuit in

---

Experiment 12–1

Connect a high-current low-voltage d.c. supply (it is a special current-limited one) to the ends of a short wire. Near the wire place another wire directly connected across the terminals of a sensitive galvanometer. Is there any deflection on the galvanometer when a current flows through the first wire, that is when the charges are moving with a uniform velocity? What happens when the current is changing, that is during switching on or off? Try replacing the direct current with a suitable alternating current. Under what conditions does the movement of charge along a wire produce a magnetic field which in turn produces an electric field?

This does not show that an electromagnetic wave is propagated—the magnetic field may not be moving fast enough.

Experiment 12–2

Connect a large capacitor in series with a 2-0-2 mA galvanometer across an inductor; this could be a large number of turns of wire on a closed iron core (Fig. 12-2). Connect a d.c. supply of a few volts across the capacitor for a short interval of time. What happens when the capacitor discharges? What happens when the capacitance in the circuit is reduced? What happens when the inductance is reduced?

an amplifier circuit. Part of the energy in the oscillatory circuit can be extracted, amplified, and then fed back into the oscillatory circuit to be in phase with the oscillations and compensate for the energy loss due to damping. A possible circuit is shown with the oscillatory circuit in the load position in the **triode amplifier** (Fig. 12–1); this gives rise to low-frequency oscillations.

Figure 12–1:

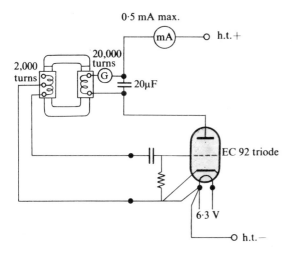

Fig. 12–1 Production of low-frequency oscillations

(Experiment 12–3.) An alternative form of oscillator is the **Hartley circuit** where the oscillatory circuit is connected between the anode and the grid of the valve. If the capacitance and inductance of the circuit are progressively reduced, the frequency of the oscillations increases.

(Experiments 12–4 and 12–5.)

(Experiment 12–6.) **Radio waves** are part of what is known as the electromagnetic spectrum and have wavelengths in the region 10^{-3} to 10^4 m. The wavelengths used in medium-band radio broadcasting are in the region 200 to 550 m. All these waves show interference, diffraction, and polarization and travel at a velocity of 3×10^8 m/s in a vacuum.

12–3 Microwaves

When you blow over the neck of an open bottle a sound is heard. This is due to the oscillation of the air molecules in the bottle. By varying the size of the air volume in the bottle the frequency of the emitted note can be varied. In a similar manner electrons can be made to oscillate in a cavity and produce very short radio waves whose frequencies are in the region of 10^{10} Hz, that is a wavelength of 3 cm. This effect is used in a valve known as a **klystron**. The radiation emitted by the

---

Experiment 12–3

Assemble the circuit shown in Fig. 12–2 and observe the oscillations produced with a meter or an oscilloscope. Change the capacitance and observe the effect on the frequency.

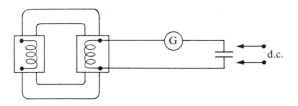

Fig. 12–2 Discharge of a capacitor through an inductor

Experiment 12–4

Try the circuits shown in Fig. 12–2 with an 0·001 μF capacitor and Fig. 12–3. The first will give oscillations in the audio-frequency region and the second in the radio-frequency region. In the first case the oscillations can be detected with a loudspeaker or oscilloscope, while in the second case the production of an e.m.f. in a single closed loop of wire can be used to show the existence of oscillations. A lamp in series with the single loop will light when the rate of change of magnetic flux through the loop is sufficiently high.

klystron can be detected by a microammeter shunted by a diode and connected to an aerial. An alternative is to feed the signal from the aerial through an amplifier to a loudspeaker. This will emit a note if the microwaves have been modulated by an audio frequency; this involves an audio-frequency wave being superimposed on the microwaves.

(Experiment 12–7.) **Microwaves** are electromagnetic waves with wavelengths of the order of a few centimetres. Because of their wave nature they can be reflected and refracted, and undergo diffraction, interference, and polarization.

Example 12–1. Which end of the radio spectrum of wavelengths is most likely to cast 'shadows' and produce areas of bad reception due to the presence of hills, etc.?

If the wavelength of a wave is comparable with the size of the object (a hill), then pronounced diffraction round the object will occur. If the wavelength is less than the size of the object, then a shadow will be produced. If the wavelength is greater than the size of the object, complete diffraction round the object will occur and the progress of the waves will hardly be affected. Thus the long-wavelength end of the radio region will give less 'shadow' than the short-wavelength end.

If in doubt try this as an experiment with different-sized objects in a ripple tank or in front of the microwave source.

Radio waves are transverse vibrations and one method of demonstrating this is to place a metal grill between the transmitter and the detector. The grill wires should be insulated from each other and less than one wavelength apart. As the grill is rotated so the detected signal varies, being a maximum when the electric field of the waves is perpendicular to the wires and a minimum when the electric field is parallel to the wires. When the electric field is parallel to the wires, oscillatory currents are set up in the wires by the electric field.

Polarization by reflection can be obtained with a plane sheet of hardboard, glass, or Perspex. Maximum reflection occurs when the electric field is parallel to plane of the reflector.

(Experiment 12–8.)

12–4 Infra-red radiation

(Experiment 12–9.) Heating effects are produced by visible radiation; the effect is greater at the red than at the violet end of the spectrum. Beyond the red end there is a much greater heating effect than in the visible part of the spectrum—the radiation in this part of the spec-

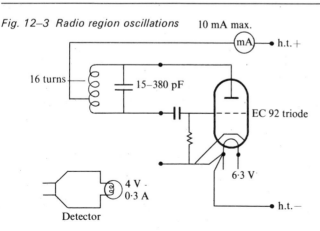

Fig. 12–3 Radio region oscillations

Experiment 12–5

Oscillations of very high frequency can be produced if the oscillatory circuit is replaced by a single loop of wire. This is equivalent to an inductance of one turn of wire with the capacitance provided by stray capacitance between the wires and in the valve. These amount to about 0·1 μH and 10 pF respectively in the case of the EC 92 valve normally used for this experiment. This gives a frequency of about 140 mHz. A suitable circuit for generating such a frequency is shown in Fig. 12–5.

The radiation can be detected by a small loop of wire and a lamp. This will not emit a very bright light because the capacitance and inductance of the detector circuit

trum is known as **infra-red radiation.** Infra-red radiation is an electromagnetic wave with wavelengths in the region between microwaves and the red end of the visible spectrum, that is about 7×10^{-7} to 10^{-4} m. The infra-red spectrum produced by dispersion in a prism does not extend very far into the infra-red owing to glass becoming opaque in the near infra-red. The spectrum can be extended out much further if glass lenses and prisms are not used—these can be replaced by curved mirrors for the focusing and production of a parallel beam and a reflection diffraction grating to produce the

dispersion. Fig. 12–4 shows the distribution of energy over the wavelengths produced by a hot solid, 'a black body', at different temperatures. A **'black body'** is one that is able to emit or absorb all wavelengths. The higher the temperature, the more the energy peak moves towards the shorter wavelengths, that is as a solid is heated the radiation goes from dull red to bright red to white.

12–5 Ultra-violet radiation

(Experiment 12–10.) The radiation beyond the violet end of the visible spectrum is known as **ultra-violet radiation.** It is an electromagnetic wave with wavelengths in the region of 4×10^{-7} to 10^{-8} m.

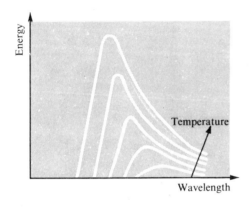

Fig. 12–4 Distribution of energy with wavelength

Fig. 12–6 Detection of radio waves

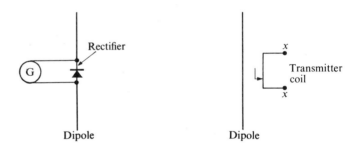

*Fig. 12–5
Production of
high-frequency
oscillations*

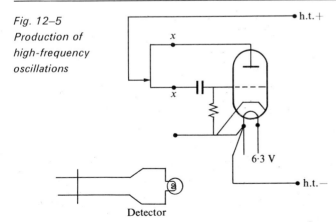

are not at the resonant values for the frequency being used, that is the detector is not correctly tuned. The detector can be tuned by varying the length of the loop —a convenient way of doing this is to add two long wires to the coil and slide a short piece of wire along them.

Replace the detector lamp by a short circuit and in place of the sliding wire move a lamp along the wires. What happens? Compare your results with those produced by waves on strings or sound waves in a tube. Do the results resemble those produced by standing waves? What is the wavelength?

12–6 X-rays

All electromagnetic waves are produced by accelerating electric charges, and one of the simplest ways of doing this is to bring electrons which are travelling at a high speed to an abrupt stop, a violent deceleration, by colliding with a solid (Fig. 12–7). The deceleration produces a radiation known as **X-rays.** X-rays can be detected by their effects on photographic plates, fluorescent screens, and the discharging of a charged electroscope. Their wavelengths are in the region of 10^{-8} to 10^{-10} m. (Experiment 12–11.)

X-rays are extremely penetrating, being able to penetrate even sheets of metal. The amount of blackening of a photographic plate depends on the intensity of the incident X-rays and hence on the density and thickness of the intervening material. Because of this X-rays can be used to detect cracks or flaws in metal castings. A crack will appear as a light line on the dark image of

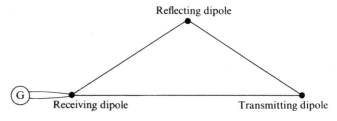

Fig. 12–8 Interference of radio waves

the casting owing to the difference in density of the metal and the air in the crack.

Interference experiments can be performed with ruled diffraction gratings, these being used at grazing incidence, that is angles of incidence close to 90°. This is necessary because the refractive index of materials for X-rays is close to unity. Polarization of X-rays can be shown by scattering from a gráphite block in much the same way that polarization of visible light can be shown by reflection.

If the wavelength–energy relationship for X-rays is investigated, a continuous spectrum of X-rays, similar

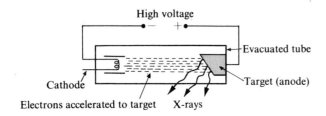

Fig. 12–7 X-ray tube

Experiment 12–6

Assemble the oscillator as in the previous experiment. In this experiment determine whether the radiation is an electromagnetic wave by looking for interference and polarization effects.

A more convenient form of the detector is to straighten out the two lengths of rod to form a vertical aerial. The length of each rod should be $\frac{1}{4}\lambda$, the resulting aerial being known as a **half-wave dipole**. The transmitter can be made more efficient by placing a half-wave dipole close to the oscillator coil and by using it as the transmitter. The best position for it relative to the coil can be found by trial and error. If the detector lamp is not bright enough, a galvanometer can be used in its place with a diode across its terminals (Fig. 12–6).

Place another half-wave dipole about 2 m from the transmitter and to one side so that it can be used as a reflector (Fig. 12–8); the detector should be about 3 m from the transmitter. What happens to the detector output as the reflector is slowly moved further away? Consider the two path lengths, transmitter to reflector to detector and transmitter direct to detector, and determine the wavelength in air of the radiation.

Try a standing-wave experiment by placing the reflector, detector, and transmitter in line, with about 3 or 4 m between the transmitter and the reflector. What

to the continuous visible spectrum produced by a hot solid, is found (Fig. 12–9). When the accelerating voltage applied to an X-ray tube is increased, the emis-

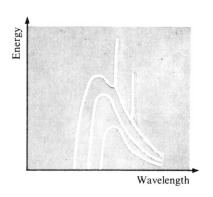

Fig. 12–9 Distribution of energy with wavelength for X-rays

sion moves to shorter wavelengths—similar to an increase in temperature for the hot solid. If high voltages are used, that is electrons of high kinetic energy, the continuous spectrum has a line spectrum superimposed on it. The continuous spectrum is the same for all solids, whereas the line spectrum is characteristic of the element of which the target is made, that is the element being hit

by the electrons. These line spectra are simpler than those produced with light in that there are fewer lines and a simple change in wavelength as we move from element to element (Fig. 12–10). The line spectra are characteristic of the atoms involved. A plot of the square root of the frequency of similarly placed lines in the spectra against the atomic number of the element yields a straight line. The various groups of lines are known as the **K, L, M,** and **N series.** The **atomic number** is the number of positive charges carried by an atom and determines the position of the element in the **periodic table.** This is a particular order of the elements dictated by their chemical properties.

Fig. 12–10 X-ray line spectra

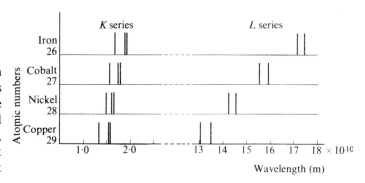

happens as the detector is moved along the line joining the transmitter and the reflector? What is the wavelength of the radiation in air?

Rotate the detector aerial when it is receiving radiation from the transmitter. What happens?

Summarize your findings concerning the nature of the radiation.

Experiment 12–7

(a) Interference. Direct the radiation from the microwave transmitter onto a double-slit arrangement with slits about 3 cm wide and 3 cm apart. Examine the radiation transmitted by the slits by moving the detector

aerial along a line parallel to the slits (Fig. 12–12) Comment on the results. What is the wavelength?

(b) Diffraction. Direct the radiation onto slits of width 3 cm, 9 cm, and 30 cm. Comment on the results. Direct the radiation onto a straight edge. Examine the 'shadow' pattern. Comment on the results.

(c) Polarization. Allow the radiation to fall directly onto the detector aerial. Rotate the detector aerial about the line joining it and the transmitter aerial. Comment on the results.

(d) Standing waves. Allow the radiation to fall normally onto a metal screen placed about 40 cm from the source. What happens as the detecting aerial is

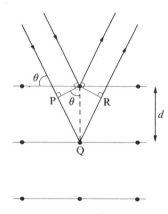

Fig. 12–11 *The reflection of X-rays from a crystal*

surface atom and one immediately beneath it in the crystal is PQ+QR. But

$$PQ = QR = d \sin \theta$$

where d is the distance between the layers of atoms. Hence for constructive interference, that is a path difference of a whole number of wavelengths,

$$2d \sin \theta = n\lambda$$
$$n = 1, 2, 3, \text{etc.}$$

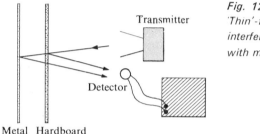

Fig. 12–13
'Thin'-film interference with microwaves

Metal   Hardboard

Though interference can be shown with a ruled diffraction grating, this is a comparatively recent event as difficulty was experienced in producing such fine rulings. At first crystals were used as diffraction gratings. Atoms in a crystal are arranged in a regular array and behave as a three-dimensional diffraction grating.

(Experiment 12–12.) Consider X-rays incident at a glancing angle θ on a layer of atoms (Fig. 12–11). Reflection will occur from atoms on the surface and from those in the various layers beneath the surface. The path difference between the ray reflected from a top

Reflection from a crystal surface only occurs at particular angles—when constructive interference occurs. At other angles destructive interference occurs. The reflection occurring when $n = 1$ is known as the first-order reflection, when $n = 2$ the second-order reflection, etc.

---

moved along the line joining the source and the screen? What is the wavelength?

(e) Reflection. Are the laws of reflection obeyed? Use a metal screen as the reflector.

(f) Refraction. Use either a paraffin wax prism or a Perspex prism filled with paraffin. Are the rays refracted?

(g) Thin films. Place a hardboard sheet in front of, and parallel to, a metal sheet. Place the detector and transmitter side by side along the normal to the sheets (Fig. 12–13). What happens as the separation of the two sheets is varied?

Summarize your results and comment on the nature of microwave radiation.

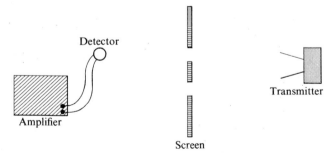

Fig. 12–12 *Interference of microwaves*

Example 12–2. What is the spacing between atoms in a crystal for which the first-order reflection occurs at a glancing angle of 30° with X-rays of wavelength 2×10^{-10} m?

$$2d \sin \theta = n\lambda$$

Hence

$$d = \frac{1 \times 2 \times 10^{-10}}{2 \times \sin 30°}$$

$$d = 2 \times 10^{-10} \text{ m}$$

X-rays can thus give information concerning the distances between atoms in crystal.

Summary

The **electromagnetic spectrum** consists of a wide range of wavelengths (Fig. 12–14), certain properties being common to all the waves. They are all produced by the acceleration of electric charges and all have the same velocity in a vacuum, 3×10^8 m/s. All are transverse waves and show interference, diffraction, and polarization effects. All can be considered as moving electric and magnetic fields.

Essentially the production of the various waves is as follows: **radio waves** by the oscillation of electrons in an electrical circuit; **microwaves** by the oscillation of electrons in a cavity; **infra-red, visible, and ultra-violet waves** by the application of energy to the outer electrons of atoms or molecules, e.g. heat; **X-rays** by the bombardment of a solid with electrons. Interference and diffraction effects are shown by slits, etc., when the opening size is of the order of the wavelength concerned. With radio waves this means openings many metres wide, while with light the openings are of the order of 10^{-4} cm, and with X-rays of the same order as the spacing between atoms in a crystal, about 10^{-10} m. Polarization effects can be shown by reflection from a suitable surface for all parts of the spectrum.

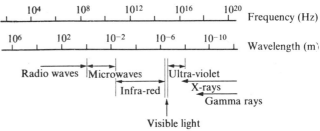

Fig. 12–14
The electromagnetic spectrum

Experiment 12–8
Investigate the polarization of microwaves by using a grid of wires or rods and also by reflection from a suitable reflector.

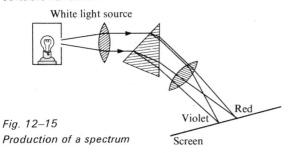

Fig. 12–15
Production of a spectrum

Experiment 12–9
Produce a parallel beam of radiation by placing a lens a suitable distance from a tungsten iodide lamp. A prism should then be placed in the beam to produce dispersion, the resulting dispersed beam being focused on a screen by a lens (Fig. 12–15). Is there anything beyond the ends of the visible spectrum? Move a blackened phototransistor (or thermopile), connected to a galvanometer, along the spectrum. What happens as the phototransistor is moved along the visible spectrum and then out beyond the ends of the spectrum?

Problems

12–1 A microwave transmitter produces 3-cm waves which are directed along the normal to a metal screen. A small detector is moved along the line joining transmitter and screen. Explain how the detector signal varies from the transmitter right up to the screen.

12–2 How are radio waves produced? (Details of circuits are not required.)

12–3 What evidence exists for believing that radio waves are waves?

*12–4 Saturn appears to behave as a black body at 90 °K. What can be said about the radiation emitted from that planet?

12–5 The lattice spacing in a calcite crystal is $3 \cdot 029 \times 10^{-10}$ m. At what angle will be the first-order reflection for X-rays of wavelength $1 \cdot 5 \times 10^{-10}$ m?

12–6 Explain how X-rays can be used for the detection of cracks and flaws in metal castings.

*12–7 Consider radio waves of wavelength 1,500 m, micro-waves of wavelength 3 cm, light of wavelength 5×10^{-7} m, and X-rays of wavelength 5×10^{-11} m each in turn incident on a cube of material of side 5 cm containing atoms a distance of 2×10^{-10} m apart. What effects would you expect to see for each type of wave?

*12–8 Imagine that you have discovered a new radiation. What properties would you look for if you wish to determine whether it is an electromagnetic wave?

12–9 What conclusions can you draw from the following experimental facts that were found when the properties of a radiation were examined.
 (a) At a particular angle of incidence no reflection was observed for the radiation incident on a block of Perspex.
 (b) Diffraction was just perceptible with slits of width $0 \cdot 1$ mm.
 (c) No deviation of the radiation beam occurred in a magnetic field.

12–10 The transverse nature of microwaves can be shown by directing them through a fine grill of wires. As the grill is rotated so the transmitted signal varies from a maximum to a minimum when the wires have been rotated through $90°$. Explain this in terms of electromagnetic induction.

Experiment 12–10

Produce the spectrum of the tungsten iodide lamp as in Experiment 12–9 but this time move a fluorescent powder or painted card along the spectrum and beyond its ends. Detergent powder can be used. What happens beyond the ends of the spectrum?

Experiment 12–11

(a) Direct the radiation from a small X-ray tube onto a charged electroscope. What happens?
(b) Direct the radiation onto a film, left in its light-proof packet. What does the developed film show? Cover part of the film, before exposure, with a piece of lead foil.

Experiment 12–12

(a) Water waves reflected from an array of points can be used to simulate the effects occurring with X-rays. The spacing of the points should be about 2λ. How does the reflection of the waves vary with the angle of incidence? Comment on the results as the angle of incidence is changed.

(b) A similar analogue experiment can be made with microwaves reflected from an array of polystyrene spheres, stacked in a manner similar to the arrangement of atoms in a crystal. Paraffin wax lenses to produce parallel radiation incident on the spheres and also to focus the reflected radiation onto the receiver improve the results. How does the reflection of the waves vary with the angle of incidence? Is the reflected intensity constant?

13 Transmission of Heat

13–1 Conduction

There are three processes by which heat can be transferred: **conduction, convection,** and **radiation.** In conduction, heat is transmitted by a body without any visible motion of the body. Thus, if you hold one end of a metal pipe and the other end is heated, your hand can detect the flow of heat by conduction along the pipe.

(Experiment 13–1.) The rate of flow of heat through materials depends on the nature of the material and on its form. Materials which contain a large amount of trapped air are good insulators. Metals are good conductors of heat. In fact materials which are good conductors of electricity are good conductors of heat.

(Experiment 13–2.) The **rate of flow of heat** along a specimen is directly proportional to the temperature difference and to the cross-sectional area, and inversely proportional to the length of the specimen. These facts can be written as

$$\frac{Q}{t} \propto \theta_1 - \theta_2$$

$$\frac{Q}{t} \propto A$$

$$\frac{Q}{t} \propto \frac{1}{L}$$

or combining these

$$\frac{Q}{t} \propto \frac{A(\theta_1 - \theta_2)}{L}$$

$$\frac{Q}{t} = \frac{KA(\theta_1 - \theta_2)}{L}$$

θ_1 is the temperature at one end of a bar specimen and θ_2 the temperature at the other end; L is the length of the bar; A is the cross-sectional area; Q is the quantity of heat passing along the bar in time t; K is the constant of proportionality and is known as the **thermal conductivity** of the material of the bar. The quantity $(\theta_1 - \theta_2)/L$ is the temperature gradient along the bar.

Experiment 13–1

Arrange a number of rods of equal diameter and length but of different materials, e.g. copper, aluminium, iron, and glass, so that all the rods are at the same temperature at one end. This can be done with sufficient accuracy if the rods are clamped on retort stands and the ends are held in the same region of a Bunsen flame. Touch the outer end of each rod in turn. Can you detect a difference from one rod to another? Write down which you think are good and which are bad conductors of heat.

Take a number of metal cans or calorimeters and stand them on corks. Wrap different materials round each can, e.g. aluminium foil, Fibreglass, and cotton wool. Try one with loosely wrapped and another with tightly wrapped cotton wool. Put hot water, at the same temperature, in each can. Measure the temperature with thermometers inserted through cardboard disks used as lids. How do the temperatures change with time? Which type of lagging material is the best heat insulator?

Experiment 13–2

The apparatus for this experiment consists of a copper bar, of 1 to 2 cm in diameter, 25 cm long, and lagged

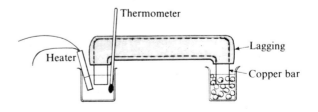

Fig. 13–1 Thermal conductivity experiment

Units:

	Q	J	cal
	t	s	s
	θ	°C	°C
	A	m²	cm²
	L	m	cm
	K	W/m degC	cal/s cm degC

Example 13–1. Determine the quantity of heat lost per second by conduction through the copper wall, 2 m² in area, of a tank when the temperature difference between the inner and outer walls is 20 degC. The copper is 1 cm thick. The thermal conductivity of copper is $3\cdot8 \times 10^2$ W/m degC.

$$\frac{Q}{t} = \frac{3\cdot8 \times 10^2 \times 2 \times 20}{0\cdot01}$$

$$= 15\cdot2 \times 10^5 \text{ J}$$

(Experiment 13–3.)

13–2 Convection

(Experiment 13–4.) The term **convection** is used when heat is transferred by the movement of matter due to changes in density. Thus, if one region of either a liquid or a gas is heated, it will become less dense than the surrounding liquid or gas. This will result in the heated substance being displaced upwards. Heat is thus transferred to other regions by movement of the fluid. Water in many small heating installations circulates by convection. A boiler heats the water which then rises through a pipe into radiators. As the water cools so it falls back through a return pipe to the boiler which is situated at the lowest point of the system.

The rate at which heat is transferred by forced convection depends on the speed of flow of the fluid. Thus, if a fan blows air over a heated object, it loses heat at a greater rate than if the air were still.

13–3 Radiation

The transfer of heat by **radiation** is the movement of energy by means of an electromagnetic wave. No material medium is necessary.

with Fibreglass pipewrap. The bar is bent at each end so that one end can dip into water in a cardboard cup and the other into melting ice in a similar cup (Fig. 13–1). When hot water is placed in one of the cups, heat will flow through the copper bar to the ice in the other cup. This will cool the hot water. The heat lost by the hot water can be replaced from a small immersion heater placed in the cup. The rate of supply of heat energy per second is $IV-J$. The cup should have a cardboard lid and the temperature of the water should be measured with a thermometer after stirring the water.

Determine the heat that must be supplied per second to maintain a temperature difference of 80 degC between the two ends of the bar. Also determine the heat input per second required for a 60-degC and a 40-degC difference. How does the heat flow per second along a bar depend on the temperature difference between its ends? The results should not be taken immediately the hot water reaches the appropriate temperature but time must be allowed for steady conditions to be reached, that is the heat input and heat losses remain constant as indicated by the temperature of the hot water remaining constant.

Repeat the experiment for the 80-degC temperature difference with bars of the same diameter but different lengths. Also try copper bars of different diameter but the same length.

What factors determine the rate of flow of heat along the bar?

Experiment 13–3
Use the results of Experiment 13–2 to obtain a value for the thermal conductivity of copper.

(Experiment 13–5.) Heat radiation is emitted by all bodies above the absolute zero of temperature. The rate at which a body emits radiation depends, however, on the nature of its surface. A dull black surface is a much better emitter than a light shiny surface. The rate of absorption of heat radiation by a body also depends on surface conditions. A dull black surface is a much better absorber than a shiny surface. A dull black surface is both a good emitter and a good absorber; a shiny surface is a poor emitter and a poor absorber. A perfect absorber and a perfect radiator is called a **black body.** The **emissive power** of a surface is the ratio of the heat energy radiated per unit surface area per unit time to that emitted per unit surface area per unit time by a black body. Clean copper has an emissive power of about 0.1×10^{-3} W/m^2, while oxidized copper has a value of about 0.7×10^{-3} W/m^2.

(Experiment 13–6.) The total energy emitted per unit time by a black body is proportional to the fourth power of the absolute temperature of the body. This is the **Stefan–Boltzmann law.** In addition to emitting radiation the body will receive radiation from its surround-ings. Thus the net loss of energy per unit area per second by radiation is

$$E = \sigma(T^4 - T_s^4)$$

where T is the absolute temperature of the body and T_s that of the surroundings. σ is known as Stefan's constant and has the value 5.7×10^{-8} W/m^2 degK^4.

Example 13–2. Determine the loss of energy per second by radiation from a surface of area 0.2 m^2 when it is at a temperature of $727\,°C$ and the surroundings are at $27\,°C$.

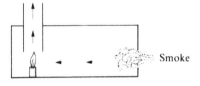

Fig. 13–2 Convection currents in air

Experiment 13–4

(a) Heat a large beaker of water over a Bunsen flame. Drop a small crystal of potassium permanganate into the water. What happens? Why?

(b) Place a lighted candle below the chimney in the model (Fig. 13–2) and allow the smoke to enter through the opening in the side of the box. What happens? Why?

Experiment 13–5

Place an electric heating element in a clamp near the top of a retort stand. Place one hand above the heater and the other hand the same distance below it. Do you detect a difference? Would you expect any difference? Explain.

Interpose a sheet of glass between the heater and your lower hand. Is there any change? What can you

deduce from your observation?

Arrange a thick sheet of copper, one side of which is blackened with soot and the other brightly polished, over a Bunsen flame. The sheet should be supported in a vertical plane by clamps and retort stands. When the sheet is hot, place your hands equal distances from each side of the plate. Do you detect a difference? Connect two identical thermopiles to separate galvanometers, and place one each side of the copper sheet and facing it. Turn off the gas supply to the Bunsen and note the galvanometer readings. Do the readings differ? If they do try to give an explanation.

Experiment 13–6

Place an electric heater vertically above a blackened thermometer, a thermopile, or a phototransistor. How does the response of the detector change as the temp-

total energy lost by the surface per second
$$= 5\cdot67 \times 10^{-8} \times 0\cdot2 \times (1{,}000^4 - 300^4)$$
$$= 1\cdot13 \times 10^4 \text{ J}$$

immersion heater to the air surrounding the tank by a combination of all three processes.
(Experiment 13–7.)

13–4 Heat transfer

Heat transfer in practice can rarely be identified as being solely due to conduction or to convection or to radiation but tends to occur by a combination of all three. Thus, in the case of a domestic hot water tank, heat is supplied to the water by an immersion heater. The heat is transferred throughout the water in the tank by convection. Near the walls of the tank there will be a layer of stagnant water and the heat will pass through this essentially by conduction. The walls will probably be coated with chemicals deposited from the water and thus conduction will again by involved. There is then conduction through the tank wall followed by conduction through a layer of stagnant air and convection in the surrounding air. In addition there will be radiation from the wall of the tank. Heat is thus transferred from the

Summary

Heat energy can be transferred by **conduction, convection,** and **radiation.** In conduction no movement of the medium occurs but a medium is necessary. With convection movement of the medium occurs. With radiation no medium is necessary as the transfer is by means of electromagnetic waves. For conduction

$$\frac{Q}{t} = \frac{KA(\theta_1 - \theta_2)}{L}$$

For radiation

$$E = \sigma T^4$$

The **rate of emission of heat** is determined by surface conditions.

erature of the heater is increased? Does the heater have to glow visibly for heat radiation to be emitted? Dull red heat is about 600 °C, bright red about 900 °C, white at about 1,500 °C. With this as a rough guide determine how the energy emitted per unit area per second varies with temperature. Use the absolute scale of temperature in your calculations.

Experiment 13–7
Examine the temperature distribution due to the passage of heat through an inside window. Thermocouples with long leads are advisable so that no person has to go near the window during the observations. Measure the air temperatures inside and outside the window, the glass temperatures, and the temperature distribution in the stagnant air on either side of the glass.

Problems

13-1 How does the rate of flow of heat through a specimen depend on the cross-sectional area of the specimen and the temperature gradient?

13-2 The thermal conductivity of copper is 3.8×10^2 W/m degC. Determine the rate at which heat is conducted along a copper bar, of cross-sectional area 1 cm^2, when one end is at a temperature of 200 °C and the other end is at 50 °C. The length of the bar is 50 cm. What assumptions must you make? Is your answer going to be too high or too low?

13-3 Describe the convection currents that arise in a beaker of water heated at one side. Why do convection currents occur?

13-4 The black-body temperature of the planets can be shown to vary as $R^{-1/2}$, where R is the distance of the Sun to the planet in astronomical units. One astronomical unit (A.U.) is the distance of the Earth to the Sun. The black-body temperature of the Earth may be taken as 277 °K.
(a) Derive the relationship $T \propto R^{-1/2}$.
(b) Calculate the black-body temperature of Mercury if it is 0.39 A.U. from the Sun.
(c) For what reasons could your answer not be the real temperature of Mercury?

13-5 How is a vacuum flask constructed and on what principles does it operate?

13-7 Why is expanded polystyrene such a good insulator?

13-8 Explain how double glazing instead of single-sheet glazing reduces the heat loss from a house.

14 Molecules

14–1 Kinetic theory of gases

Gases exert pressure; the Earth's atmosphere exerts a pressure capable of supporting a column of mercury 76 cm high. Why does a gas exert pressure? We might consider a gas to be a collection of randomly moving particles and their bombardment of the walls of the container or surfaces placed in the gas as causing the pressure. Pressure is force per unit area and when a particle strikes a surface a force is experienced by the surface. Consider what happens if somebody throws a weight at you—you certainly notice if it hits you.

(Experiments 14–1 and 14–2.) The Brownian motion of smoke particles offers evidence for the random motion of gas molecules. In this case smoke particles are struck by air molecules and are caused to move randomly. Their motion is the result of a difference in the number of collisions occurring on each side of the ash particle.

When the volume of a gas changes, the pressure changes and, in the case of gases like dry air, there is a simple relationship between the pressure and volume if the mass and temperature of a gas remain constant. This is known as **Boyle's law:**

$$PV = \text{a constant}$$

(Experiment 14–3.) When the temperature of a gas such as air is changed there is a change in volume if the pressure is constant or, if the volume is constant, there is a change in pressure. These can be summarized in the equations

$$V/T = \text{a constant} \qquad \text{if } P \text{ is constant}$$
$$P/T = \text{a constant} \qquad \text{if } V \text{ is constant}$$

Together with Boyle's law these give the general equation

$$PV/T = \text{a constant}$$

T is the absolute temperature.

A word of warning in applying these equations—consider a gas of water molecules at a temperature above 100 °C and the consequence of a reduction in temperature; when 100 °C is reached liquefaction occurs and the gas laws certainly do not apply to a liquid; thus the gas laws would not be expected to apply to any gas all the way down to absolute zero as liquefaction would affect the results.

Experiment 14–1

Place a smoke cell on the stage of a microscope. This is a small cell in which smoke can be introduced from a piece of smouldering rope and the smoke particles observed by a microscope. What happens to the smoke particles? Can the effects be explained on the basis of a gas, in this case air, behaving as though it consisted of a large number of small particles moving around in a random manner?

The motion of the smoke particles is known as Brownian motion.

Experiment 14–2

Adjust the flexible base of the kinetic theory model so that it is at the top of the stroke of the vibrator. Place the ball bearings in the apparatus (Fig. 14–1) and set the vibrator in motion. This causes the bearings to be set in motion. Consider this as a model of a gas. Introduce a polystyrene foam sphere among the ball bearings. Is the motion of the polystyrene sphere similar to that of the smoke particles in the previous experiment?

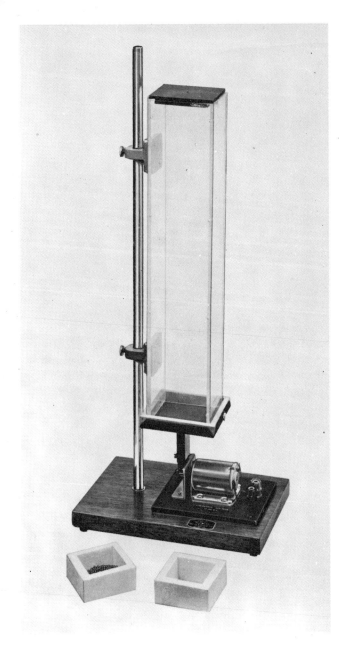

Fig. 14–1 Kinetic theory apparatus

(Experiments 14–4 and 14–5.) The force on the wall of a container of gas is due to molecular bombardment. According to Newton's laws the force experienced by a surface due to a collision is the rate of change of momentum. Linear momentum is the product of the mass and velocity of the colliding particle. The total force experienced by an area of wall is the product of the total number of molecules hitting the wall and the rate of change of momentum of a single molecule.

Example 14–1. (a) What is the momentum of a ball of mass $\frac{1}{2}$ kg moving with a velocity of 4 m/s?

$$\text{linear momentum} = \text{mass} \times \text{velocity}$$
$$= \tfrac{1}{2} \times 4 = 2 \text{ kg m/s}$$

(b) If this ball hits a wall and stops dead, what is the change of linear momentum?

The momentum after collision is zero. The momentum before collision is 2 kg m/s. Hence the change of momentum of the ball is 2 kg m/s. Momentum of the wall before the collision is zero. Momentum of the wall and whatever it is attached to after the collision is 2 kg m/s. This must be so if momentum is to be conserved. Hence the change of momentum of the wall is 2 kg m/s. This must be so if the ball stops.

Experiment 14–3

Place a polystyrene or cardboard disk in the kinetic theory model so that the ball bearings move inside the enclosure. With no motion of the ball bearings the volume of the enclosure will be zero as the disk slides down to the bottom. Set the vibrator in motion with a constant applied voltage and mark on the side of the container the volume produced. Why does the disk move? Add another equal-mass disk so the 'gas' now supports twice the force spread over the area of the disk, that is twice the pressure. What is the volume? Try more disks. How is the pressure related to the volume?

What happens if the voltage is changed? What

(c) If 100 of these balls hit the wall, what is the total momentum gained by the wall?

total momentum gained
$= 100 \times 2 = 200$ kg m/s

(d) What happens to the wall if we had a ball stuck to the wall and then suddenly it jumped away from the wall with a momentum of 2 kg m/s?

The wall must obviously experience an opposite momentum change if momentum is to be conserved, that is the wall receives a momentum of 2 kg m/s.

(e) What is the momentum gained by the wall if 100 balls jump away from the wall, each having a momentum of 2 kg m/s?

total momentum gained by the wall
$= 100 \times 2 = 200$ kg m/s

(f) What is the momentum gained by the wall if 100 balls each moving with a speed of 4 m/s and having a mass of $\frac{1}{2}$ kg hit the wall and rebound with the same speed? (The momentum gained by the wall when the balls are stopped and when they move off the wall are in the same direction.)

total momentum gained by the wall
$= 100 \times \frac{1}{2} \times 4 + 100 \times \frac{1}{2} \times 4$
$= 400$ kg m/s

(g) If these balls take 5 s to hit the wall and rebound, that is this is the time taken for the momentum to change by 400 kg m/s, calculate the force experienced by the wall.

Force is rate of change of momentum, in this case 400 kg m/s, in 5 s. Hence the force experienced by the wall is 400/5 kg m/s^2 or 80 N.

(h) If the area of the wall is 0·5 m^2, what is the pressure on the wall?

Pressure is force per unit area and is thus

$80/0·5 = 160$ N/m^2

The previous example outlines a way to determine the pressure on the wall of a container due to molecular bombardment. Consider molecules moving in a box. The momentum of a single molecule is mv and the force experienced by a wall of the box when a molecule hits it and rebounds is the rate of change of momentum. The change of momentum for this one molecule is $2mv$ if it

happens to the energy of the ball bearings when the voltage is changed? How does the energy of the ball bearings affect the results at constant pressure?

Experiment 14–4
Reduce the number of ball bearings in the gas model so that some estimate of the number striking the cardboard disk can be made. How does the number of collisions change when the pressure is doubled?

Experiment 14–5
Drop ball bearings one at a time onto the pan of a lever balance. Does the balance pan experience a force due to

the collisions? In both this and the previous experiment how can the force be made constant, that is the pressure constant? The pressure produced by a gas in Boyle's law apparatus is not constantly fluctuating.

Experiment 14–6
Determine the average speed of air molecules at room temperature by measuring the mass and pressure of a known volume and using the equation $PV - \frac{1}{3}Nm\overline{v}^2$. Nm is the mass of gas of volume V.

Measure the volume of a large and light, but strong, container full of air at room temperature and atmospheric pressure. Weigh the container. Then connect the

rebounds with the same speed. But how many collisions does this molecule make with the walls per second? If the box is of length L and the molecule is bouncing between opposite walls (Fig. 14–2), then the distance travelled in time t is vt and the number of collisions made with one wall is $vt/2L$. Thus the total change of momentum in t seconds for the molecule at some wall of the box is $2mv(vt/2L)$. Hence the force exerted on the wall is $2mv(vt/2L) \div t$ or mv^2/L. The pressure on the wall is this force divided by the area of the wall A, that is mv^2/LA. But LA is the volume V of the box, hence the pressure is mv^2/V. This is the pressure due to just a single molecule. If there are N molecules in the box and they move randomly, it is reasonable to assume, for large numbers, that there will be $\frac{1}{3}$ of the molecules colliding with the wall we have considered. Hence the total pressure is $\frac{1}{3}Nmv^2/V$. Thus

$$PV = \tfrac{1}{3}Nmv^2$$

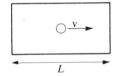

Fig. 14–2 *Molecule bouncing between walls of a container*

We have assumed that the speed v is the same for all the molecules, but some molecules will have lower and some higher speeds. The speeds will vary owing to collisions. Thus we take the average value of v^2, denoted by $\bar{v}^2$.

Now, if N, m, and $\bar{v}^2$ are constants and if the average kinetic energy is proportional to the temperature of the gas, the expression $PV = \frac{1}{3}Nm\bar{v}^2$ will give Boyle's law, this is $PV = $ constant.

(Experiment 14–6.)

The speed of air molecules at room temperature is of the order of 500 m/s or 1,000 mile/h. This is only the average speed; some molecules will have a higher speed and others a lower.

If the temperature of a gas is directly proportional to the average kinetic energy of the molecules, then all gases at the same temperature would be expected to have the same average kinetic energy. Thus in air the average kinetic energy of oxygen molecules will be the same as the average kinetic energy of nitrogen molecules.

Example 14–2. How are the speeds of oxygen, v, and nitrogen molecules, V, in air related to each other? The mass of the oxygen molecule is 16 units by comparison with nitrogen at 14 units.

container to a filter pump. Start the pump and reweigh the container when the greater part of the air in it has been removed—hence obtain the approximate mass of air that occupied that volume. Determine the atmospheric pressure using a barometer and hence obtain a value for the average molecular speed.

An alternative method is to increase the pressure in the container by means of a bicycle pump and measure the increase in weight produced by the extra air. The volume of the extra air at atmospheric pressure can be obtained by releasing the air under water into a rectangular container. The volume of air released in bringing the container back to atmospheric pressure can be determined by measuring the height of the air column above the water.

kinetic energy of the oxygen molecules
$$= \tfrac{1}{2} mv^2 = \tfrac{1}{2} \times 16 \times v^2$$

kinetic energy of the nitrogen molecules
$$= \tfrac{1}{2} \times 14 \times V^2$$

As both gases are at the same temperature

$$\tfrac{1}{2} \times 16 \times v^2 = \tfrac{1}{2} \times 14 \times V^2$$

Hence

$$v^2 = \frac{14}{16} V^2$$
$$v = 0.94\, V$$

The ratio of the masses of the oxygen to the nitrogen molecules is in fact the ratio of the densities of the gases at the same temperature and pressure.

density of oxygen at 0 °C and 76 cmHg pressure
$= 1.43$ g/l

density of nitrogen under the same conditions
$= 1.25$ g/l

ratio of the densities $= 1.43/1.25 = 16/14$

Example 14–3. Calculate the speed of the oxygen molecules at 0 °C and 76 cmHg pressure.

density of oxygen under these conditions $= 1.43$ g/l

density of mercury at 0 °C $= 13.6$ g/cm^3

$$PV = \tfrac{1}{3} Nmv^2$$

density $d = Nm/v$

Hence

$$P = \tfrac{1}{3} dv^2$$

density $d = 1.43$ g/l $= 1.43 \times 10^{-3}$ kg/l
$$= 1.43 \times 10^{-3} \times 10^3 \text{ kg/m}^3$$

The pressure in units of force per unit area, newtons per square metre (N/m^2), is given by $P = HDg$, where D is the density of mercury $= 13.6$ g/cm$^3 = 13.6 \times 10^{-3} \times 10^6$ kg/m^3, and H is the height of column of mercury $= 76$ cm $= 76/100$ m.

$$P = \frac{76}{100} \times 13.60 \times 10^3 \times 9.81$$
$$= 1.01 \times 10^5 \text{ N/m}^2$$

Hence

$$v^2 = \frac{3P}{d} = 3 \times \frac{1.01 \times 10^5}{1.43}$$
$$v = 466 \text{ m/s}$$

If the pressure of a gas is varied at constant temperature, the volume must vary also.

$$PV = \text{a constant}$$

Such a change is known as an **isothermal change,** a change at constant temperature, and to be isothermal must be carried out very slowly. When isothermal conditions are required, the average kinetic energy of the molecules must not change during the change in volume. If a change in kinetic energy occurs, there will be a change in temperature.

(Experiment 14–7.)

Example 14–4. A quantity of gas is expanded isothermally to four times its original volume under isothermal conditions. By what factor does the pressure change?

$$PV = \text{a constant}$$
$$P_1 V_1 = P_2 4 V_1$$
$$P_1/P_2 = 4$$

The final pressure is one quarter of the initial pressure.

When a reduction in volume takes place slowly, the heat produced by the reduction is dissipated to the surroundings and no noticeable change in temperature of the gas occurs—with a fast reduction in volume the heat cannot be dissipated quickly enough and the temperature rises. Such a change of volume is called an **adia-** **batic change.** The temperature change produced in a quantity of matter by a certain amount of heat will depend on the specific heat of the material. Thus, when we change the volume of a gas adiabatically, that is when the heat is not dissipated to the surroundings, we should expect the resulting pressure change to be determined by the specific heat; the pressure also depends on the volume and the temperature.

$$PV/T = \text{a constant}$$

Thus in addition to the normal (P, V, T) relationship there will be an equation linking the changes with the specific heat. In the case of a gas there are two principal specific heats: one when heat is supplied at constant volume, and another when the heat is supplied at constant pressure. The heat energy in the first case produces a change in molecular kinetic energy, while in the second case the energy is used to produce a molecular kinetic energy change and a change in volume, that is external work is done. For an adiabatic change

$$PV^\gamma = \text{a constant}$$

where γ is the ratio of the specific heat at constant pressure to the specific heat at constant volume.

Example 14–5. Consider Example 14–4 when a gas expanded to four times its original volume but now let

---

Experiment 14–7
With the kinetic theory model is there any difference in the behaviour of the 'molecules' if a change in volume occurs slowly or quickly?

the change be adiabatic. By what factor does the pressure change? Take the ratio of specific heats as 1·4.

$$P_1 V_1^{1·4} = P_2 (4V)^{1·4}$$
$$P_1/P_2 = 4^{1·4} = 6·96$$

(Experiment 14–8.)

When a ball hits a stationary surface, its speed of rebound is almost the same as its incident speed. If, however, the surface is moving, e.g. a cricket bat, then the ball will rebound with a different speed. If the bat and ball are moving towards each other, the ball moves faster after the collision than before; if the bat and ball are both moving in the same direction, the ball will move more slowly after the collision. If a molecule hits a surface which is moving towards it, then it rebounds with a higher velocity, and thus, as the gas temperature is proportional to the average kinetic energy of the molecules, there is an increase in temperature. If the surface is moving away from the molecule, there will be a drop in temperature after the collision.

Example 14–6. Calculate the temperature change when gas in a container has its volume suddenly increased by a factor of ten. The initial temperature is 300 °K and the ratio of specific heats is 1·3.

$$PV/T = \text{a constant} \qquad PV^{\gamma} = \text{a constant}$$

Eliminating P between these two equations gives

$$TV^{\gamma-1} = \text{a constant}$$

Therefore

$$T_2 V_2^{\gamma-1} = T_1 V_1^{\gamma-1}$$
$$T_2 = T_1 \frac{V_1}{V_2} = 300 \left(\frac{1}{10}\right)^{0·3}$$
$$\approx 300 \times \frac{1}{2·0}$$
$$\approx 150 \,°\text{K}$$

Hence there is a temperature drop. This is one method used to produce low temperatures.

14–2 Molecular size

How big is a molecule, what is its mass, how many are there in a given volume, and how far do they move between collisions with each other?

The size of a molecule may be a meaningless question if molecules are not hard spheres. The answer would depend on how hard the molecules were 'squeezed'. One technique for obtaining an effective size is to allow one type of molecule to collide with another;

Experiment 14–8
Blow up a balloon, with a thermocouple inside it (Fig. 14–3). What happens when (a) the air is slowly released, and (b) the air is rapidly released? In both cases the change in volume and pressure is the same. The pressure can be measured with a simple water manometer.

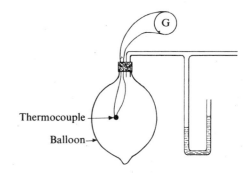

Fig. 14–3 *Experiment on adiabatic changes*

the distance travelled between collisions will depend on the sizes of the molecules.

(Experiments 14–9 and 14–10.) The motion of bromine molecules through air can be considered to involve a very large number of collisions. In the absence of air the bromine molecules travel very fast; thus in air we can consider the bromine molecules to move very rapidly between collisions with air molecules. The average distance between collisions is known as the **mean free path.**

(Experiment 14–11.) The motion of the bromine molecule through air is similar to the motion of the coloured marble through the other marbles. In Experiment 14–10 we obtained the average distance travelled by bromine molecules in 500 s—this is the result of a lot of collisions by the bromine molecules. How is this average distance related to the mean free path?

(Experiment 14–12.) In a random walk the average distance travelled in a straight line from the original position is that due to $\sqrt{N}$ steps, where N is the number of steps taken. Thus for a molecule taking 25 steps the average distance travelled from the original position is 5 steps. In Experiment 14–10 the average distance travelled by a bromine molecule in 500 s was measured. Hence it is possible to calculate the number of steps taken by the molecule.

Example 14–7. Consider the average distance travelled by a bromine molecule in air in 500 s to be 10 cm. What is the mean free path?

Let L be the mean free path. This is the average length of step. Hence the average distance travelled is $L\sqrt{N}$; N is the total number of steps. Thus

$$0 \cdot 1 = L\sqrt{N}$$

The total distance travelled in 500 s is the product of the speed of the molecule and the time. At room temperature bromine molecules have an average speed of 200 m/s. Thus

$$\text{total distance travelled} = 500 \times 200 = 100,000 \text{ m}$$

The number of collisions made in that time is

$$N = 100,000/L$$

Hence we can determine both N and L.

$$0 \cdot 10 = \frac{100,000}{N}\sqrt{N}$$

$$0 \cdot 10 = \frac{100,000}{\sqrt{N}} = \frac{10^5}{\sqrt{N}}$$

$$(0 \cdot 10)^2 = \frac{10^{10}}{N}$$

Experiment 14–9
Place two wagons with a magnet fixed to each on a short length of model railway track. The two magnets should be arranged to repel each other. Push one truck gently along the line towards the other one which should be stationary. What is the effective size of the trucks if we consider contact to have been made when the second truck begins to move?

Experiment 14–10
This experiment uses liquid bromine—bromine is dangerous, it attacks most things and, if any is spilt on

the skin, blisters are liable to occur. The vapour also is dangerous and should not be inhaled. If any splashes occur, they should immediately have a strong ammonia solution poured on them. When the apparatus is being cleaned after the experiment it should be done in a bucket of dilute ammonia solution with the person handling the apparatus using rubber gloves.

(a) Observe the motion of bromine molecules in a vacuum. The bromine capsule, which has very thin walls, should be introduced into a thin-wall section of rubber tubing which is attached to the main tube via the Interkey stopcock (Fig. 14–4). Before the rubber tubing and ampoule are attached, the

Thus

$$N = 10^{12}$$

This is the number of collisions made in 500 s. Hence

$$L = \frac{100,000}{10^{12}}$$

$$L = 10^{-7} \text{ m}$$

The mean free path is the average distance travelled by a molecule between collisions—this will obviously depend on the size of the molecule and their density, e.g. if you were running through a crowd of people all standing in a definite area, your mean free path would depend on the number of people in the crowd and their size.

For two molecules to collide, their centres must be the sum of the two radii apart (Fig. 14–5), that is $r_1 + r_2$. In travelling the mean free path only one collision, on the average, occurs. In travelling the distance L, collisions will occur with all the molecules in a volume $\pi (r_1 + r_2)^2 L$. But, if L is the mean free path, only one molecule exists in this volume. Thus in the gas the number of molecules per unit volume is

$$\frac{N}{V} = \frac{1}{\pi (r_1 + r_2)^2 L}$$

or

$$L = \frac{1}{\pi (r_1 + r_2)^2 N/V}$$

The value of N/V depends on the pressure of the gas; $PV = \frac{1}{3}Nmv^2$. Hence

$$L \propto 1/P$$

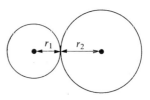

Fig. 14–5 Collision of two spheres

Example 14–8. If the mean free path of an air molecule is 10^{-7} m at atmospheric pressure, what is the value when the pressure is reduced by a factor of 10^7?

As $L \propto 1/P$ then the mean free path at this low pressure is 1 m.

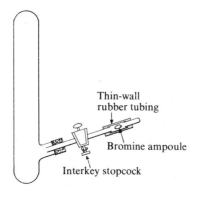

Fig. 14–4 Bromine diffusion apparatus

Thin-wall rubber tubing

Bromine ampoule

Interkey stopcock

main tube should be evacuated using a vacuum pump. The bromine is released into the vacuum by squeezing the rubber section containing the ampoule and so breaking it. How long does it take for the bromine molecules to move through the full length of the main tube? On the basis of kinetic theory is the result as expected?

(b) Observe the motion of bromine molecules in air. This is a repeat of the previous experiment without the main tube being evacuated. Do the bromine molecules diffuse through the air molecules as fast as they move through a vacuum? How far does the average bromine molecule travel in air in 500 s?

This is in fact the order of magnitude of the pressure in a cathode-ray tube or electronic valve.

In order to use our knowledge of the mean free path to arrive at the radius of a molecule we need to know N/V. The density of a gas is Nm/V. This equation applies equally to liquids and solids. Thus the ratio of the density of a liquid to the same material in the gaseous state is the ratio of the number of molecules per unit volume in the liquid to that in the gas. The density of liquid nitrogen is approximately 750 times that of gaseous nitrogen at room temperature. If we consider a nitrogen molecule to occupy a small cube of side equal to the diameter of the molecule when in the liquid, then the number of molecules per unit volume in the liquid is one per d^3, d being the diameter of the molecule. Thus in the gas the volume occupied by one molecule is $750d^3$. Hence the number per unit volume is

$$1/750d^3$$

Now using our mean-free-path equation and assuming that the radius of a bromine molecule is near enough that of a nitrogen molecule,

$$L = \frac{1}{\pi d^2 N/V}$$

$$L = 10^{-7} = \frac{750d^3}{\pi d^2}$$

Hence the diameter of an air molecule is approximately 4×10^{-10} m. The volume occupied by one nitrogen molecule at atmospheric pressure is $750d^3$ or $750 \times (4 \times 10^{-10})^3$ m^3. Thus the number of nitrogen molecules per cubic metre is

$$\frac{1}{750 \times (4 \times 10^{-10})^3} \approx 2 \times 10^{25}$$

Example 14–9. How many molecules are there in a television tube of volume 0.25 m^3 and at a pressure lower than atmospheric by a factor of 10^7?

The mean free path of an air molecule at this pressure was determined in the previous example as 1 m. Hence the number of molecules per unit volume can be found from the mean-free-path equation.

$$L = \frac{1}{\pi d^2 N/V}$$

$$L = \frac{1}{\pi (4 \times 10^{-10})^2 N/V}$$

Hence

$$N/V \approx 2 \times 10^{18}$$

Experiment 14–11

Put a number of marbles in a tray, one of them being a different colour to the rest. Agitate the tray and observe the motion of the differently coloured marble.

Experiment 14–12

Draw on a piece of paper six lines radiating from a point each making 60° with the next, or use a 60° triangular grid marked paper. Mark the six directions 1, 2, 3, 4, 5, and 6. Starting from the middle of the paper throw a die to indicate the direction your molecule moves and move it 1 cm in that direction. Repeat this for a number of

steps by the molecule, starting each step where the previous one finished. Do this for 25 steps. How far is the final position of the molecule from the start? How is this distance related to the total distance covered? Compare your results with those obtained by other members in the class.

Experiment 14–13

Direct a water stream along a horizontal glass tube along which three vertical tubes have been placed to measure the pressure (Fig. 14–6). How does the pressure vary along the tube? Now repeat the experiment with a tube which has a narrow middle section

Thus the number of molecules in the tube, volume 0.25 m^3, is 5×10^{17}. (As the number of molecules per unit volume is directly proportional to the pressure, we could have directly deduced that reducing the pressure by a factor of 10^7 would reduce the number per unit volume by the same amount.)

Knowing the number of molecules per unit volume in air and the density of air we can obtain a value for the mass of an air molecule.

$$\text{density} = 1.2 \text{ kg/m}^3$$

Hence

$$1.2 = mN = 2 \times 10^{25} \times m$$

Thus the mass of an air molecule is of the order of 6×10^{-26} kg.

14–3 Low pressure

Pressure is force per unit area and thus has units of N/m^2. In many cases, however, it is more convenient to refer to pressures in terms of the height of a column of mercury that the pressure will support. Thus we have pressure units of **millimetres of mercury** (mmHg). This unit is sometimes known as a **torr.** The **standard atmospheric pressure** is 760 mmHg. The pressure at the foot of a column of mercury of this height is the force per unit area (Fig. 14–8), that is the weight of the column divided by the area of the base of the column:

$$\text{pressure} = hAdg/A$$

where A is the cross-sectional area, d is the density of the mercury, h is the height of the column, and g is the acceleration due to gravity. The pressure is thus hdg.

Atmospheric pressure supports a column of mercury 760 mm high and this is the pressure at room temperature produced by the collisions of 2×10^{25} molecules/m^3 on the walls of a container or in the case of the atmosphere on the surface of the Earth. By removing molecules we reduce the pressure.

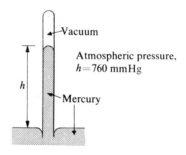

Fig. 14–8 Simple barometer

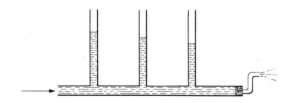

Fig. 14–6 Fluid flow along a uniform tube

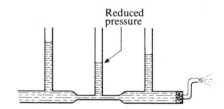

Fig. 14–7 Fluid flow through a constriction

(Fig. 14–7). How does the pressure vary along the tube? What is the effect of the narrower section? What happens if the flow rate is increased?

(Experiment 14–13.) When a fluid flows through a constriction a pressure drop occurs. This is known as the **Bernoulli effect.** This occurs because, for the same volume of fluid to flow per second through both the narrow and the wide tubes, the fluid velocity in the narrow tube must be greater than that in the wide tube. Because a change in speed occurs, there is an acceleration and hence there must be a force applied. This force is applied by a difference in pressure between the two tubes. Thus the pressure in the narrow tube is lower than that in the wide tube. If, for example, in a **water-jet pump** the flow rate is high enough and the constriction small enough, air is sucked, through the middle tube, into the water stream. Hence the device can be used to extract air from a vessel and so reduce the pressure in the vessel. The apparatus generally takes the form shown (Fig. 14–9). This water-jet pump is capable of reducing the pressure from atmospheric down to a few millimetres of mercury.

The water-jet pump is not capable of producing pressures lower than a few millimetres, so a **rotary pump** is used in which air is swept out of a vessel by a rotating vane (Fig. 14–10). The rotating solid cylinder is set off centre but contact is maintained with the walls by a pair of spring-loaded vanes. As the cylinder rotates, the vanes trap some of the air, and sweep it out through the

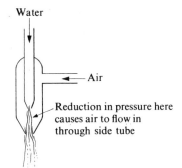

Fig. 14–9 Water-jet pump

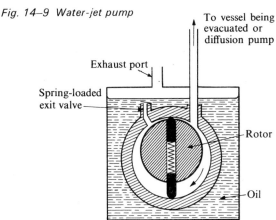

Fig. 14–10 Rotary pump

exit port. The entire system is immersed in oil which lubricates the system and also forms a seal between the low- and high-pressure sides of the pump. Such a pump can reduce the pressure to as low as 10^{-3} mmHg.

For still lower pressures the **diffusion pump** is used. This pump produces a high-velocity stream of oil molecules which forces the air molecules by a series of collisions towards the pump exit (Fig. 14–11). The oil is heated electrically and after ascending the central chimney the oil molecules are deflected downwards by fixed vanes. This downward moving stream traps air molecules diffusing into it, and sweeps them along to the rotary pump, while the oil vapour condenses on the cold walls of the pump and runs down to the bottom.

A diffusion pump cannot operate at atmospheric pressure; it is essentially a low-pressure device and must therefore operate through another pump capable of producing a pressure of about 10^{-1} mmHg, that is a rotary pump. Pressures as low as 10^{-7} mmHg can be produced with a diffusion pump. The normal operating procedure is to use the rotary pump to reduce the pressure to 10^{-1} mmHg or less and then start the diffusion pump.

To measure pressures from about 10^{-1} to 10^{-4} mmHg a device depending on Boyle's law can be used. In this method a fixed volume of gas at the low pressure

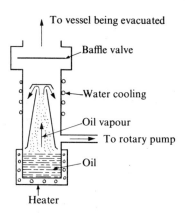

Fig. 14–11 Oil diffusion pump

is taken and compressed to a smaller volume and this new volume and higher pressure are then measured. The instrument is known as a **McLeod gauge** (Fig. 14–12). The mercury reservoir is lowered until the bulb is connected, via the branch at B, to the vessel being evacuated, that is the vessel in which we require to know the pressure. The gas is the vessel to be evacuated and in the bulb is then at the same pressure P_1. The mercury reservoir is now raised, so trapping a volume V_1 of air

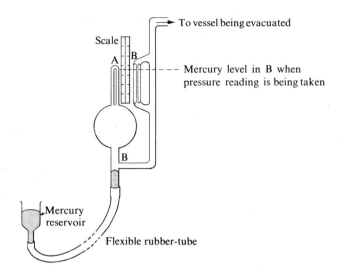

To vessel being evacuated

Scale

A B

Mercury level in B when
pressure reading is being taken

B

Mercury
reservoir

Flexible rubber-tube

Fig. 14–12 McLeod gauge

$$P_1 V_1 = P_2 V_2$$

But

$$V_2 \doteqdot hx \qquad \text{and} \qquad P_2 = h$$

h is the difference in heights of the mercury in A and C, and x is the cross-sectional area of the capillary bore. Therefore

$$P_1 = \frac{P_2 V_2}{V_1} = \frac{xh^2}{V_1}$$

Hence, if V_1 and x are known and h is measured, the pressure P_1 in the vessel being evacuated can be found.

Another type of gauge for use at these pressures is the **Pirani gauge.** This is based on the fact that the temperature of a hot wire depends on the rate at which it loses heat to the surrounding gas. Quite a large amount of heat is lost by the wire owing to air molecules striking and rebounding with a higher velocity, having acquired energy in the collision. The number of molecules available for collision will therefore determine part of the heat loss. The temperature of the wire is therefore related to the gas pressure. The temperature does, however, affect the resistance of the wire. The wire, contained in a glass envelope, is therefore inserted in one arm of a Wheatstone bridge (Fig. 14–13) and the out-of-balance current taken as a measure of the pressure.

in the bulb and the capillary A at the pressure P_1. Raising the mercury further compresses the air in A to a new pressure P_2 in a new volume V_2. The mercury also enters the capillary C and its height is adjusted until it is level with the inside of the top of the tube A. Then

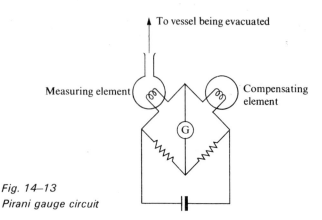

Fig. 14–13
Pirani gauge circuit

To compensate for room-temperature changes a sealed gauge not connected to the vacuum vessel is placed in the opposite arm of the bridge. This type of gauge has to be calibrated and can be used over the range 10^{-1} to 10^{-4} mmHg.

Low pressures are essential to the working of certain equipment. For example, electronic valves and cathode-ray tubes, through which electrons need to travel more or less unimpeded by air molecules require low pressures. The evaporation of metals onto surfaces to form mirrors is done in a low-pressure chamber. The pressure in the chamber is reduced until the mean free path of the metal

atoms is greater than the distance from their source to the mirror (Fig. 14–14). Another major use of low pressures is in the freeze drying of certain foods. At a low pressure, foodstuffs rapidly loose their water content and a drop in temperature occurs owing to the high rate of evaporation. The final product can be easily stored and then reconstituted in its original form by the addition of water.

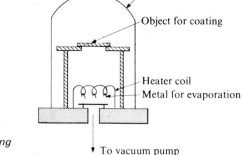

Fig. 14–14
Apparatus for
vacuum coating

14–4 The Earth's atmosphere
The Earth's atmosphere exerts a pressure equal to that of a column of mercury about 760 mm high. This pressure is due to the motion of air molecules, that is molecules of oxygen, nitrogen, carbon dioxide, and water vapour together with other gases in lesser pro-

---

Experiment 14–14
Use the kinetic theory model without a disk in the tube. Examine the 'gas' in the tube when the vibrator is in motion and the volume of the gas is only restricted by gravity. Is there a uniform density?

Experiment 14–15
With the aid of two thermometers construct a wet and dry bulb hygrometer. Determine the two temperatures and with the aid of tables determine the relative humidity of the room.

Experiment 14–16
Take a polythene bag, large enough for you to put your hand in it and leave some space spare. Place calcium chloride or some other water absorber in it. Cut one end of the bag so that the wet and dry bulb hygrometer can be inserted (Fig. 14–15). Allow steady conditions to be reached —after a few minutes —and obtain a reading for the relative humidity in the bag. Then insert your hand. What happens to the thermometer readings? Why? Is your hand comfortable?

Repeat the experiment with the air in the bag saturated with water vapour. This can be done by removing the drying agent and replacing it with water.

portions. The density of the Earth's atmosphere varies with height, being greatest at the lower levels; thus the atmospheric pressure varies with height.

(Experiment 14–14.)

Although water vapour is only a very small percentage of the Earth's atmosphere, it is a very important factor in our existence. The amount of water vapour in a particular locality varies with time, from hour to hour, and determines personal comfort as well as weather conditions. The vital factor, however, is not how much water vapour there is in the air but how much it can hold. If the air is saturated with water vapour, then condensation (rain) can occur. In addition a saturated atmosphere produces discomfort as evaporation from the human body is reduced. Heat is continually being generated in the human body and it must be dissipated if the body temperature is to remain constant. This dissipation is assisted by the evaporation of perspiration from the skin. In saturated air no evaporation of perspiration will occur and so the body temperature will rise.

The moisture content of air relative to that necessary to saturate it is known as the **relative humidity:**

$$\text{relative humidity} = \frac{\text{mass present in a certain volume}}{\text{mass to saturate the same volume at the same temperature}}$$

The relative humidity is generally expressed as a percentage. An alternative method of expressing relative humidity is

$$\frac{\text{actual pressure of water vapour present}}{\text{saturation vapour pressure of water at the same temperature}}$$

The **saturation vapour pressure** is the vapour pressure when surplus liquid is present and no more liquid can be vaporized. This assumes that the vapour pressure is proportional to the mass of water present.

Instruments used to measure the relative humidity are known as **hygrometers.** The most direct method is to use an absorption hygrometer in which air is passed over a drying agent such as calcium chloride and to determine the change in weight for a certain volume of air. The experiment is then repeated with air that has been drawn through water and is therefore saturated. A more convenient method is to use a wet and dry bulb hygrometer which depends on the amount of cooling experienced by a body when water evaporates from it. The evaporation rate will be the greatest when the relative humidity is the lowest, and hence, as the rate of evaporation determines the cooling produced, a temperature measurement can be used as a measure of the relative humidity. The temperature drop also depends on the room temperature, so allowance must be made for this.

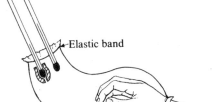

Calcium chloride   Elastic band

Fig. 14–15

Experiment 14–17

Cool a test-tube by placing in it ether (take care—ether is an anaesthetic) and bubbling air through it. An alternative method is to put a mixture of water and ice in the tube. A thermometer should be placed in the tube so that the temperature of the tube can be measured. Observe the outside of the tube as the temperature falls. What happens?

Note the temperature at which dew is deposited on the tube.

The instrument consists of two thermometers, one of which is as normal and reads the room temperature, while the other has its bulb sheathed in a layer of muslin which dips into water. The water evaporating from the muslin causes a drop in temperature. The instrument must be shielded from draughts.

(Experiments 14–15 and 14–16.)

When air is cooled, the mass of water vapour necessary to saturate the air becomes less and at some particular temperature, the **dew point,** saturation occurs and excess water becomes deposited on surfaces. The saturation vapour pressure at the dew point is equal to the vapour pressure at the original temperature and thus

relative humidity

$$= \frac{\text{saturation vapour pressure at the dew point}}{\text{saturation vapour pressure at the original air temperature}}$$

Tables of saturation vapour pressures at different temperatures are available.

(Experiment 14–17.)

Example 14–10. Determine the relative humidity given that the air temperature is 20 °C and the dew point is 5 °C.

saturation vapour pressure at 20 °C = 17·5 mmHg and at 5 °C = 6·5 mmHg

Hence

$$\text{relative humidity} = \frac{6·5}{17·5} \times 100$$

$$= 37\%$$

There are many other forms of hygrometer—a common form is the hair hygrometer, based on the fact that the length of a human hair depends on the relative humidity. A hair will lengthen by about 3% when the relative humidity increases from a low value to saturation. The hair is made to actuate a pointer giving a reading on a scale. The device requires frequent calibration.

Summary

A gas can be considered as composed of molecules, rather like ball bearings, moving around in a random manner within a container. Pressure on the walls is due to molecular bombardment. By consideration of the collisions on the walls a relationship involving the pressure can be obtained

$$PV = \tfrac{1}{3} Nm\overline{v^2}$$

The temperature of the gas is considered to be proportional to the average kinetic energy of the molecules:

$$T \propto \tfrac{1}{2} m\overline{v^2}$$

These equations in fact yield the **gas laws:**

$$PV = \text{a constant} \qquad \frac{V}{T} = \text{a constant} \qquad \frac{P}{T} = \text{a constant}$$

These combine to give

$$\frac{PV}{T} = \text{a constant}$$

An **isothermal change** is one which takes place at constant temperature, and thus Boyle's law applies. An **adiabatic change** is one which takes place when heat neither enters nor leaves the gas and for this condition

$$PV = \text{a constant}$$

Typical speeds for air molecules at room temperature are of the order of 500 m/s, with the molecules being about 4×10^{-10} m in diameter and having a mass of 6×10^{-26} kg.

Low pressures can be produced in the range of centimetres of mercury by using the **Bernoulli effect** (a pressure drop occurs when a fluid increases in speed owing to a constriction in a tube), in the 10^{-3} mmHg range by using a **rotary pump** which sweeps the gas out of a container by mechanical means, and in the 10^{-6} mmHg range by the use of a **diffusion pump.** In this a stream of oil molecules is used to carry the air molecules away from the vessel being evacuated. The **McLeod** gauge can be used down to pressures of about 10^{-3} mmHg and depends on Boyle's law. A **Pirani gauge**, depending on heat being removed from a hot filament, by the surrounding gas measures down to about 10^{-4} mmHg.

The **relative humidity** of air is defined as

$$\frac{\text{mass of water vapour present in a certain volume}}{\text{mass needed to saturate the same volume at the same temperature}}$$

or

$$\frac{\text{actual vapour pressure of water present}}{\text{saturated vapour pressure of water at the same temperature}}$$

The relative humidity can be measured by a wet and dry bulb hygrometer. Another method is to determine the temperature at which dew forms.

relative humidity

$$= \frac{\text{saturation vapour pressure at the dew point}}{\text{saturation vapour pressure at the original air temperature}}$$

Problems

14-1 The molecules in a gas are considered to be in a continual state of random motion. What evidence is there for this statement?

14-2 How can ball bearings bouncing about in a container be used as a model for a gas?

14-3 According to the gas equation the volume of a gas is directly proportional to the absolute temperature of the gas. What would you expect to happen as the temperature is reduced?

14-4 How does the average speed of a gas molecule change as the temperature is changed?

*14-5 What would be the effect on the equation $PV = \frac{1}{3}Nm\overline{v^2}$

of only a small number of molecules being present in the volume considered?

14-6 Calculate the speed of hydrogen molecules at $0\,°C$ and 76 cmHg pressure if the density of hydrogen under these conditions is 0.090 g/l. The density of mercury at $0\,°C$ is 13.6 g/cm³.

14-7 What change in temperature is produced when a gas has its volume suddenly increased by a factor of four? The initial temperature of the gas may be taken as $300\,°K$. The ratio of the specific heats for the gas is $1:3$.

*14-8 How could you define a collision on a molecular scale?

*14-9 In the derivation of the gas equation $PV = \frac{1}{3}Nm\overline{v^2}$ from kinetic theory does the shape of the molecule considered have any significance?

14-10 A gas at $27\,°C$ is allowed to expand (a) isothermally, and (b) adiabatically from 2 l to 4 l. If the initial pressure was 2 atm, calculate in both cases the final pressure and temperature. The ratio of specific heats of the gas may be taken as $1:4$.

14-11 The mean free path of helium molecules when measured at 76 cmHg pressure and $0\,°C$ is 28×10^{-8} m. What is their cross-sectional area under these conditions?

14-12 The effective diameter of an oxygen molecule is 3.5×10^{-10} m at $0\,°C$ and 76 cmHg pressure. What percentage of the space is occupied by the gas molecules?

14-13 Why does a pressure drop occur when a fluid flows from a wide tube to a narrower tube?

14-14 What is the vapour pressure of water present in the atmosphere when the relative humidity is 60%? The s.v.p. at the temperature concerned is 20 mmHg.

14-15 Explain how dew is produced.

*14-16 The mean free path of air molecules in a tube is 20 cm. If the tube is 15 cm long and only 4 cm in diameter does this mean that molecules do not collide with each other in this tube?

15 Electrons

15–1 Thermionic emission

(Experiment 15–1.) When heat energy is supplied to a liquid, evaporation occurs—when heat energy is supplied to a solid, electrons, having negative charge, can be emitted. The electron current emitted from the cathode of an electronic valve depends on the temperature of the cathode and the material from which it is made. Because the emitted electrons are negative, making the collector electrode (anode) positive causes a current to flow through the diode, while making it negative prevents a current flowing. If the anode is made sufficiently positive, saturation is reached when the anode is collecting the electrons as fast as they are produced (Fig. 15–2). When saturation occurs no change in current can be produced by any further increase in anode potential.

The **diode valve** (Experiment 15–1) allows current to flow only when the anode is positive with respect to the cathode. When the polarity is reversed, no current flows. Therefore, when an alternating voltage is applied across the diode, current flows only during the half-cycle when the anode is positive with respect to the cathode, and during the rest of the cycle no current flows. If the applied voltage is sinusoidal, then the current wave consists only of half-sine waves. If a suitable resistor is connected in series with the anode and the positive terminal of the h.t. supply, an oscilloscope connected across the resistor will show a half-wave voltage corresponding to the current.

(Experiment 15–2.)

15–2 Deflection of charged particles by electric and magnetic fields

When the anode is positive with respect to the cathode in a diode, electrons will accelerate from the cathode to the anode. The acceleration is produced by the electric field between the two electrodes.

electric field strength $E = V/d$

where V is the potential difference, and d is the distance over which the electrons accelerate, that is the distance between the electrodes. The force exerted on a charge by an electric field is

force $F = Eq$

where q is the charge.

Experiment 15–1

For this experiment a large demonstration diode and a sensitive galvanometer are required. In the diode one electrode is known as the cathode, in this case a single-wire filament, and the other the anode, a flat plate adjacent to the cathode (Fig. 15–1). Connect one side of the cathode and the anode to the terminals of the galvanometer. Is any current detected?

Heat the cathode by connecting a 4-V supply across it. What happens? Try 5 V and then 6 V. Does this make any difference?

Make the anode negative by connecting it to the negative terminal of a d.c. high-tension power supply

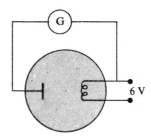

Fig. 15–1 Demonstration diode circuit

(0 V) and the cathode positive by connecting it via the galvanometer to the positive terminal. What happens as

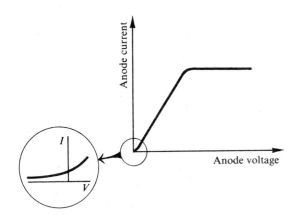

Fig. 15-2 Current–potential-difference characteristic for a diode

initial velocity is zero, then the velocity at the anode is

velocity $v = (2ad)^{1/2}$

(This equation can be obtained by using the straight-line motion equations of $v = at$ and $d = \frac{1}{2}at^2$ and eliminating t. See Chapter 1.)

$$v = \left(\frac{2Vq}{m}\right)^{1/2}$$

Units:
force	newtons (N)
mass	kilogrammes (kg)
distance	metres (m)
charge	coulombs (C)
potential difference	volts (V)
field strength	volts per metre (V/m)
velocity	metres per second (m/s)

This force causes an acceleration which according to Newton's laws is

acceleration = force/mass

$a = Eq/m$

$a = Vq/md$

m is the mass of the charged particle. The acceleration over a distance d results in a change in velocity. If the

Example 15–1. Calculate the velocity with which electrons strike the anode in a diode if a d.c. potential difference of 100 V is applied between cathode and anode. Take the charge-to-mass ratio for the electron as $1 \cdot 77 \times 10^{11}$ C/kg.

Hence

$v = (2 \times 100 \times 1 \cdot 77 \times 10^{11})^{1/2}$
$\quad = 6 \times 10^6$ m/s

---

the anode is made more negative? Reverse the polarity of the h.t. supply. What happens as the anode is made more positive? What is being emitted from the filament when it is heated?

Experiment 15–2
Plot a graph showing how the current passed by a diode depends on the potential difference applied between anode and cathode.

What is the resistance of the valve when the anode is (a) negative, and (b) positive but below saturation? What happens if an a.c. potential is applied between

anode and cathode instead of a d.c. potential? This can be observed using an oscilloscope (Figs 15–3 and 15–4)

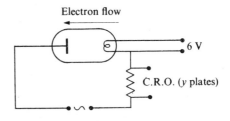

Fig. 15–3 Rectification circuit

If the electric field direction is at an angle to a beam of charged particles, then a force will be exerted on the beam in the direction of the electric field. The electric field can be produced between two parallel flat deflector plates by a potential difference maintained between them. The beam of charged particles can be produced by accelerating the charged particles (electrons) produced at a cathode to an anode and allowing them to pass through a hole in the anode. Beyond the anode the accelerating electric field is effectively zero and so electrons continue to travel with the velocity they have acquired in reaching the anode.

(Experiment 15–3.)

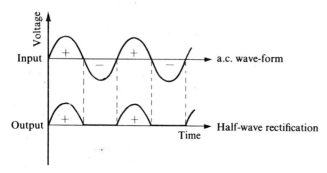

Fig. 15–4 Current wave-forms before and after rectification

A current-carrying conductor in a magnetic field is acted upon by a force; the force, the current, and the component of the magnetic field responsible are all at right angles to each other. Since a beam of electrons is a current, this too is subjected to a force when in a magnetic field. In the case of a wire, the force is given by

$$\text{force} = BIL$$

where I is the current, B is the component of the magnetic flux density at right angles to the wire, and L is the length of the wire. Current is the rate of movement of charge and thus the beam current is the rate at which the electrons move. Consider two points a distance L apart along the beam. If the electrons have a velocity v, then in time L/v all the electrons between the two points considered will have passed the second point. If there are n electrons in the beam between these points, the current is

$$nq/t$$

where q is the charge of an electron, and t is L/v. Hence the current is nqv/L.

$$\text{force} = BIL$$
$$= BnqvL/L$$

Experiment 15–3

The **Teltron deflection tube** (Fig. 15–5) contains a filament and a perforated anode to give a beam of electrons, two deflector plates, and a screen. A flat beam of electrons is produced and, where it strikes the screen, fluorescence occurs, revealing the position of the beam.

Connect a 6-V supply to the filament and connect the anode to the positive terminal of an e.h.t. supply, the negative terminal being connected to one side of the cathode. Gradually increase the e.h.t. until the impact position of the beam on the screen is visible. What happens to this trace when an electric field is applied at right angles to the beam by means of the

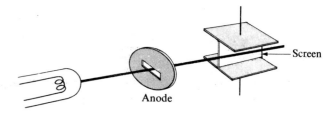

Fig. 15–5 Deflection tube details

deflector plates? The plates will require a d.c. potential of about 2 kV; this is about the same as applied to the anode and the same supply can be used for both.

Therefore the force is $Bnqv$. The force on a single electron is therefore Bqv.

Units: flux density   webers per square metre (Wb/m^2)

      charge   coulombs (C)
      velocity   metres per second (m/s)
      force   newtons (N)

When a moving object experiences a force at right angles to its direction of motion, it follows a circular path. Thus the path of a beam of electrons in a magnetic field which is at right angles to the beam is a circular path at right angles to the magnetic field. The relationship between the force and the radius of the path for an object moving with speed v is

$$\text{force} = mv^2/r$$

m is the mass of the object, and r is the radius of the path. Hence for a single electron

$$Bqv = mv^2/r \qquad (15\text{–}1)$$

As both the velocity and the magnetic flux density can be determined, a measurement of r enables the charge-to-mass ratio for the electron to be determined. The velocity can be determined from the potential difference applied between anode and cathode:

$$v = \left(\frac{2Vq}{m}\right)^{1/2} \qquad (15\text{–}2)$$

Hence eliminating v between Eqns (15–1) and (15–2)

$$Bq = \frac{m}{r}\left(\frac{2Vq}{m}\right)^{1/2}$$

$$\frac{q}{m} = \frac{2V}{B^2 r^2}$$

The magnetic flux density B either can be measured using a fluxmeter or can be calculated from the current in the coil and the coil dimensions.

The value of the charge-to-mass ratio for electrons travelling at these comparatively low velocities has been found to be $1{\cdot}77 \times 10^{11}$ C/kg.

Example 15–2. Calculate the radius of the path for electrons which have been accelerated through 200 V into a magnetic field of flux density 2×10^{-4} Wb/m^2.

$$r^2 = \frac{2 \times 200}{1{\cdot}77 \times 10^{11} \times 2^2 \times 10^{-8}}$$

$$r = 0{\cdot}76 \text{ m}$$

(Experiment 15–4.)

Remove the potential difference from the deflector plates and bring a bar magnet close to the tube. (Be careful not to touch any of the electrical connections with the magnet.) What happens? In place of the bar magnet use two coils, connected in series and placed one on each side of the tube, so that their magnetic fields both act in the same direction. The direction of the magnetic field is along the axis of the two coils. At what angle to the beam must the magnetic field be to give maximum deflection of the trace on the screen? Compare your observation with that for the force on a current-carrying conductor in a magnetic field.

Experiment 15–4
Use the Teltron deflection tube (Experiment 15–3) to obtain a value for the charge-to-mass ratio for the electron. The flux density for the coil arrangement used with this apparatus is given by

$$B = \frac{8\mu_0 Ni}{5\,(5)^{1/2}R}$$

N is the number of turns per coil, μ_0 is a constant, $4\pi \times 10^{-7}$ H/m, known as the absolute permeability of free space, i is the coil current, and R is the radius of the coils. Is there any change in the charge-to-mass ratio if the speed of the electrons is changed?

15–3 The charge on the electron

Beam deflection methods yield values of the charge-to-mass ratio but not independent values of either the charge or the mass. In the **Millikan method** for the determination of the charge on the electron very small oil drops are produced from an atomizer and fall through a hole in the upper of two parallel plates (Fig. 15–6). Most of the drops will be electrostatically charged owing to friction. An electric field is maintained between the plates and adjusted until a charged drop remains stationary—the force of gravity on the drop being balanced by the force due to the electric field.

$$Mg = QE = Q\frac{V}{d}$$

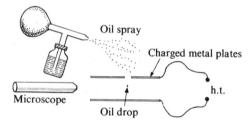

Fig. 15–6 Outline of Millikan's apparatus

Oil spray

Charged metal plates

Microscope

Oil drop

h.t.

where M is the mass of the oil drop, g the acceleration due to gravity, E the electric field strength, V the potential difference between the plates, d the distance between the plates, and Q the charge on the oil drop. The mass of the oil drop can be found from a measurement of the density of the oil and the radius of the drop:

$$M = \tfrac{4}{3}\pi\rho r^3$$

ρ is the density of the oil, and r is the radius of the oil drop. The density is found by a conventional method—the radius, however, is found by determining the terminal velocity of the drop when falling between the plates in the absence of the electric field.

According to Stokes's law the viscous resistance experienced by a sphere falling in a fluid is $6\pi\eta rv$, where η is the coefficient of viscosity of the fluid, in this case air, r is the radius of the sphere, the oil drop, and v is the terminal velocity. When these constant velocity conditions occur, the forces acting on the drop balance:

$$Mg = 6\pi\eta rv + \text{upthrust}$$

The upthrust is due to the weight of air displaced by the drop and is equal to $4\pi r^3\sigma g/3$, where σ is the density of air, but the upthrust is small enough to be neglected. When a drop starts to fall from rest, it accelerates owing

Experiment 15–5

Determine the electrostatic charge on a number of oil drops with Millikan's apparatus. The plates will require a voltage supply of up to 300 V. The manufacturer's instructions should be followed for the particular apparatus used. Generally these will consist of the following.

(a) Level the instrument.

(b) Pass a fine wire through the hole in the upper plate; focus on the wire with the microscope. Both the microscope scale and the wire should be in focus.

(c) Produce a fine spray of drops above the hole in the upper plate.

to its weight *mg*. As its speed increases, the viscous resistance $6\pi\eta r v$ increases. When the forces acting on the drop cancel, it moves with a constant velocity—remember Newton's laws. Thus in the equation $Mg = Q(V/d)$ the only unknown is Q.

(Experiment 15–5.)

Examine the results of Experiment 15–5. Are the charges carried by the oil drops continuously variable or does charge exist in well-defined quantities? For example, milk could be bought in well-defined quantities, pint bottles, or in continuously variable amounts if bought loose. Thus, if a large amount of milk were considered, it would be possible to decide whether it consisted of a number of well-defined quantities or not. If it did, then the large amount would always be a whole number of pint bottles, e.g. 23, 16, 14 pints, etc., while, if the milk could exist in any quantity, the result would in general be continuously variable, e.g. 20·34, 16·75 pints, etc. From the results of the experiment it will be found that, whatever amount of charge exists on an oil drop, it is always equal to a whole number times a basic small quantity of charge. This is the smallest possible charge and is the charge on the electron.

The charge on the electron by this and other methods is always found to be $1·6 \times 10^{-19}$ C. The result is the same regardless of whether the electrons are produced by thermionic emission or by other means. The charge on the electron is the fundamental quantity of charge. Combining the results for the charge-to-mass ratio and the charge value leads to a mass value for the electron of $9·1 \times 10^{-31}$ kg.

15–4 Electron optics

When light passes from one medium to another of different refractive index it is refracted, this property being used to produce converging or diverging beams by a suitable choice of the shape of the boundary between the two media concerned, e.g. a converging or diverging lens. Electrons when moving at an angle to an electric or magnetic field are deflected. By a suitable design of the electrodes, or the magnet, electron beams can be made either converging or diverging.

Electrons will always try to move along the direction of the electric field—thus in the experiment with the deflection tube the electron beam tends towards the direction of the electric field (Fig. 15–7). The electric lines of force are at right angles to the lines of constant potential (**equipotentials**).

(Experiment 15–6.)

Fig. 15–8 shows electron lenses with the electron paths and the equipotentials marked. In the examples

(d) Adjust the potential difference between the plates until one drop remains stationary in the field of view. Note the potential difference.

(e) By varying the applied potential difference move the drop up near the top of the scale. Then remove the potential and allow the drop to fall freely. Time the drop over the scale distance and hence calculate the velocity. (The drop reaches terminal velocity in a very short distance of fall.)

(f) Measure the density of the oil with a hydrometer or use the figure supplied by the manufacturer. The separation of the plates will generally be supplied by the manufacturer.

(g) Calculate the charge carried by the oil drop. Repeat the experiment to determine the charges carried by a number of drops.

Experiment 15–6

Plot the equipotential lines for an electric field system produced by a constant potential difference between two parallel plates. One method is to use a graphite-coated paper (recorder paper) and either paint the electrodes on it with electrically conducting paint or attach metal strip electrodes to it. A 12-V d.c. potential difference should then be applied between the electrodes. The lines of equipotential can be found by

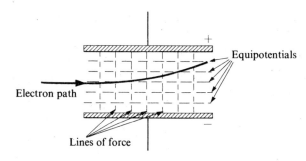

Fig. 15–7 *Equipotentials and lines of force between parallel plane electrodes*

shown the electrons are accelerating as they move through the system and tend to be less affected by the direction of the electric field as they progress.

Electron lenses can be used in the **electron micro-**

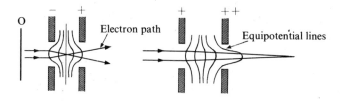

Fig. 15–8 *Electron lenses*

scope. The electron beam can be focused and produces images in an analogous manner to light rays and has the advantage of enabling much higher magnifications to be reached. However, to avoid the use of high voltages, magnetic lenses in the form of specially designed coils are more generally used.

In the **cathode-ray oscilloscope** (Fig. 15–10) a beam of electrons is brought to a focus on a fluorescent screen. Before reaching the screen the beam passes through two sets of deflector plates. The pairs of plates are set at right angles to each other and, when potential differences are applied, give deflections of the beam in the horizontal, or x axis, and in the vertical, or y axis, directions. When a steady potential difference is applied to the x plates, the spot on the screen is displaced in the horizontal direction, the displacement of the spot being proportional to the applied potential difference. Similarly, if a potential difference is applied to the y plates, a deflection proportional to the potential difference is produced in the vertical direction. If an alternating potential difference is applied to the y or x plates, the spot is drawn out into a line, half the length of which is proportional to the peak value of the applied potential difference. If a potential difference whose magnitude increased linearly with respect to time is applied to the x plates and a varying signal is applied

connecting a voltmeter between one of the electrodes and a probe and then by finding the positions of the probe point on paper which give constant-voltage readings (Fig. 15–9). Thus a two-dimensional picture of the equipotentials in three dimensions will be obtained. Examine the equipotential system between two parallel disk electrodes maintained at different potentials and both containing holes for the electron beam to pass through.

Experiment 15–7

Adjust the brightness and focusing on an oscilloscope to bring the spot into a clear focus on the screen. The

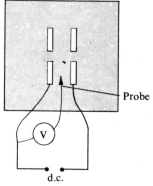

Fig. 15–9 *Equipotential plotting*

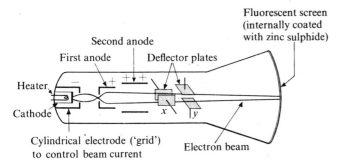

Fig. 15–10 Cathode-ray tube

electric current at any particular voltage is directly proportional to the intensity of illumination. A determination of the charge-to-mass ratio for these electrons yields the same value as for electrons produced by thermionic emission.
(Experiment 15–9.)

The production of current by the cell depends also on the colour of the light, e.g. the caesium cell gives no response to red light but is quite sensitive to yellow or green, while a potassium cell does not respond to yellow or red but is sensitive to blue.

to the y plates, then the spot traces out the variation of the y-plate signal with respect to time.
(Experiment 15–7.)

15–5 Photoelectric emission
(Experiment 15–8.) When heat energy is supplied to a solid, it is possible to eject electrons; similarly, when light energy is supplied to a solid, it is also possible for electrons to be ejected. This process is known as **photoelectric emission.** The current–voltage graph for a photoelectric cell (Fig. 15–11) is similar to that for the diode, the difference being that instead of a number of lines for different heater temperatures there are lines for different intensities of illumination. The photo-

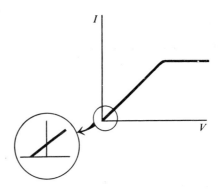

Fig. 15–11 Current–potential-difference characteristic for a vacuum photocell

brightness control alters the number of electrons striking the screen per unit time (that is the beam current) by varying the potential on the sleeve electrode surrounding the cathode; the more negative the sleeve relative to the cathode, the fewer are the electrons which can get through the hole in the sleeve and pass down the tube to the screen. The focusing is generally adjusted by changing the potential on the first anode, that is by altering the 'shape' of the electric field.

With the time base disconnected, apply a d.c. potential difference to the x and y plates in turn. Then apply a d.c. potential difference to the x and y plates at the same time. What does the spot indicate? Now apply an a.c. potential difference to the x and y plates in turn. Then apply a.c. potential differences to the x and y plates at the same time. What does the trace show?

Apply the time base to the x plates and an a.c. signal to the y plates. What does the trace show?

Experiment 15–8
Connect a spot galvanometer across the terminals of a photoelectric cell; one terminal is connected to a layer of caesium, the cathode, and the other is connected to a central rod electrode, the anode. What happens when light shines on the caesium?

15–6 Other forms of electron emission

When an electron travelling at high speed strikes a surface, it may case the emission of several other electrons from that surface. This is known as **secondary emission** and is used in the photomultiplier to produce comparatively large currents from a few primary electrons. The photoelectrons produced from a layer of material, the photocathode, such as caesium by the incident light (Fig. 15–12) are accelerated to an electrode called a **dynode** which is maintained at a positive potential with respect to the caesium layer. The electrons strike this surface and cause secondary emission—

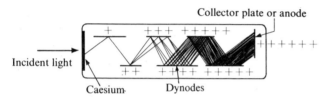

Incident light

Collector plate or anode

Caesium   Dynodes

Fig. 15–12 Principle of photomultiplier tube

thus, for each incident electron, possibly four electrons may be emitted. These electrons are then accelerated to yet another dynode, more positive than the previous,

where further secondary emission occurs. Thus after two secondary emissions there are $16(4^2)$ times more electrons than produced at the cathode. After n such electrodes there are 4^n electrons, for each electron emitted at the caesium.

Example 15–3. How many stages must there be in a photomultiplier tube for an amplification of one million to be obtained? The secondary emission coefficient, that is the ratio of the number of emitted electrons to the number incident, for the dynodes can be taken as 4.

$$\text{amplification} = 10^6 = 4^n$$

Hence

$$n = 10$$

Emission of electrons can also be produced by the presence of intense electric fields (of the order 10^8 N/C) at the surface of a material **(field emission).**

15–7 Electrons in solids

Energy is necessary to cause the evaporation of a liquid and the amount necessary to cause a definite mass to escape is known as the latent heat—there is a definite

---

Make the caesium positive with respect to the rod electrode by introducing a variable h.t. supply (Fig. 15–13). What happens as the caesium is made more positive? Try making the caesium negative—what happens?

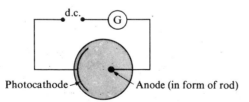

d.c.   G

Photocathode   Anode (in form of rod)

Fig. 15–13 Photoelectric cell

Experiment 15–9
Determine the current–voltage characteristics for a photocell. Take readings for different intensities of illumination. Because of the variation of the intensity of illumination with the inverse square of the distance, the intensity of illumination at the cell can be varied in a known manner by varying the separation of the cell and lamp by known amounts, e.g. doubling the distance reduces the intensity by a factor of four. Determine the relationship between the intensity and the current at some particular voltage.

amount of energy necessary for a molecule to escape. Similarly a definite amount of energy must be supplied before an electron can escape from a solid. The **work function** is the amount of energy that must be supplied to an electron to overcome surface forces and cause emission. Thus, if E is the supplied energy and ϕ the work function, then the maximum energy an emitted electron can have is

$$E - \phi$$

Electron energies are usually measured in electron volts, one electron volt (eV) being the energy acquired by an electron when accelerated through a potential difference of one volt. Hence 1 eV is 1.6×10^{-19} J. Caesium, a typical material used in photoelectric cells, has a work function of 1.8 eV, and tungsten, generally used as the filament material in lamps, a work function of 4.5 eV. These values are significantly affected by surface impurities.

Example 15–4. What is the maximum energy of emission of electrons from the caesium cathode in a photocell when the incident energy is 3 eV? The work function of caesium is 1.8 eV.

$$\text{maximum energy of emitted electrons} = 3 - 1.8 \text{ eV}$$
$$= 1.2 \text{ eV}$$

15–8 Light and electron emission

The maximum kinetic energy of electrons can be determined from a knowledge of the potential difference that must be applied to stop the electrons. Thus in a photocell, if the anode must be made 1 V negative with respect to the cathode in order to reduce the electron current to zero, then the electrons have a maximum energy of 1 eV.

(Experiment 15–10.) It would seem reasonable to expect that the higher the intensity of illumination the greater the energy of the ejected electrons—this, however, is not true. The energy of the photoelectrons is not related to the intensity of illumination; the only effect of increased intensity is to increase the number of electrons liberated, that is to increase the current. The energy of the electrons is, however, directly proportional to the frequency of the light.

$$E \propto f$$
$$E = hf$$

h is known as **Planck's constant**, 6.62×10^{-34} J s. This can be obtained from the results of Experiment 15–10.

What is the significance of these results? For a water wave the energy is not related to the frequency but depends on the wave amplitude. In photoelectricity

---

Experiment 15–10

A Mullard 90AV photocell is to be exposed to different intensities of illumination and different wavelengths (colours) of light and the maximum energy of the emitted electrons measured.

(a) Find the effect of different intensities of illumination.

A convenient piece of apparatus is shown in Fig. 15–14. Light falls on a photoelectric cell; the colour of the light can be chosen by putting filters between the lamp and the cell, and the intensity can be changed by using one, then two, then three, etc., filters of the same colour together.

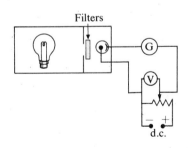

Fig. 15–14 Details for the photoelectricity experiment

light would not appear to behave like a water wave. This is astonishing considering that in interference and diffraction light behaves in a similar manner to water waves. The only possible explanation is that in photoelectricity light behaves like particles and not like waves. The energy of light particles, or **photons,** we can consider to be proportional to the frequency (colour) of the light. The intensity of illumination at a surface can depend on the number of light particles received per unit time. Thus, if the surface is coated with a photoelectric material (e.g. caesium), the energy of the emitted electrons will depend on the energy associated with the light particles, that is the frequency and the number emitted per unit time will depend on the light intensity.

Example 15–5. Calculate the maximum energy of the electrons emitted when light of wavelength (a) 8×10^{-7} m, (b) 5×10^{-7} m, and (c) 3×10^{-7} m is incident on a caesium surface of work function 1·8 eV.

Planck's constant $= 6·62 \times 10^{-34}$ J s

velocity of light $= 3·0 \times 10^{8}$ m/s

incident energy $= hf = \dfrac{hc}{\lambda}$

c is the velocity of light, and λ is the wavelength.

Hence

$$\frac{hc}{\lambda} = E' + \phi$$

E' is the maximum energy of the emitted electrons. Thus, for a wavelength of 8×10^{-7} m,

$$\frac{hc}{\lambda} = \frac{6·62 \times 10^{-34} \times 3·0 \times 10^{8}}{8 \times 10^{-7}}$$

$$= 2·49 \times 10^{-19} \text{ J}$$

$$= \frac{2·49 \times 10^{-19}}{1·6 \times 10^{-19}} \text{ eV}$$

$$= 1·56 \text{ eV}$$

Therefore

$$E' = 1·56 - 1·8 \text{ eV}$$

Emission cannot occur as the light is of insufficient energy. For a wavelength of 5×10^{-7} m, $hc/\lambda = 2·49$ eV and thus $E' = 0·69$ eV. For a wavelength of 3×10^{-7} m, $hc/\lambda = 4·16$ eV and thus $E' = 2·36$ eV.

15–9 Electron collisions with atoms

With all the electron tubes so far considered it has been assumed that the electrons do not meet any obstruction

Measure the voltage, negative with respect to the cathode, that must be applied to the anode to reduce the current to zero for different intensities of illumination. A galvanometer capable of detecting currents of the order of 10^{-7} A should be used.

(b) Measure the negative potential which must be applied to the anode to reduce the current to zero when different-colour light is used.

What is the relationship between the maximum energy of the electrons and the intensity of illumination? Is there any change in energy of the electrons when the colour of the light is changed? Plot a graph between the maximum electron energy and the light frequency. Is the light energy and hence the electron energy related to the intensity or the frequency of the light?

Experiment 15–11

The **Neva-controlled excitation tube** contains a cathode to produce electrons, a wire mesh called a grid to control the electron flow, and a collector plate (anode) (Fig. 15–15). The number of electrons is determined by the cathode current and their energies by the potential difference maintained between the cathode and the grid. A measure of the electron energy at a particular cathode current can be obtained by connecting a galvanometer between the grid and the anode. If the

in their travels between electrodes. What happens when an electron hits an atom? In any collision energy will be conserved. In a perfectly elastic collision the energy both before and after the collision is purely kinetic, while for an inelastic collision the initial kinetic energy is partly converted into another form of energy—the kinetic energy after the collision is not equal to the kinetic energy before.

Example 15–6. Are the following collisions elastic or inelastic?

(a) Two table tennis balls collide and no permanent deformation or heat is produced.

(b) Two pieces of Plasticine collide and there is no resulting motion after the collision.

In (a) kinetic energy is conserved and the collision is therefore elastic. In (b) kinetic energy is not conserved and the collision is therefore inelastic. The energy is used in changing the shape of the Plasticine and also dissipated as heat.

(Experiment 15–11.) As the accelerating voltage is increased for the valve containing gas, so the anode current rises until at some particular value of voltage there is a sharp drop in current (Fig. 15–16). After

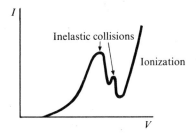

Fig. 15–16 Current–potential-difference curve obtained in Experiment 15–11

reaching a minimum the current again rises until another sharp drop occurs at a higher value of voltage than previously. After a number of sharp drops the current eventually rises very steeply and the experiment has to stop to avoid damaging the tube. When the current drops, inelastic collisions must be occurring as the electrons have lost energy and not so many reach the collector. When inelastic collisions occur, some of the kinetic energy of the electrons is taken up by the gas atoms and emitted in the form of light; consequently there is a drop in current.

The atoms in a gas are only able to accept certain energy values and they re-emit this energy in the form

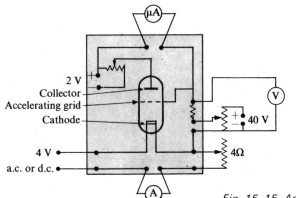

anode is given a small negative potential with respect to the grid, the lower-energy electrons will not be collected.

Measure the current between the anode and the grid as the accelerating potential is gradually increased. Provided no kinetic energy is lost by the electrons on their way through the gas in the tube, a mixture of mercury vapour and neon, the collector current should rise as the accelerating potential is increased.

When kinetic energy is lost by electrons, what happens to the gas atoms? Observe the gas with a spectroscope during the experiment.

Fig. 15–15 Apparatus for the controlled excitation of spectra

The photographs show the spectral lines that appear when mercury vapor is bombarded with electrons. In each case the bombarding electrons have a definite energy which is recorded at the side of the photograph. (Photos from: John A. Eldridge, "The Spectrum of Mercury Below Ionization," *Physical Review*, Vol. 23, Series 2, 1924.)

(a)
7.0 ev

(a) With electrons of any energy above 4.9 ev, the strong ultraviolet line of wavelength 2537 angstroms is emitted. This corresponds, as shown at the right, to an energy change in the atom from the first excited state to the ground state. By the time the energy is 7 ev, there is also an energy change from 6.67 ev to the ground state, giving a line of wavelength 1849 angstroms. The wavelength of this light, however, is too short to affect the photographic plate. Here this transition is shown by a dotted arrow. In the subsequent figures that arrow is omitted.

(b)
8.4 ev

(b) At 8.4 ev three new lines of longer wavelength show up in the region photographed. The photons of this light have energy given by the transitions from the levels around 7.8 ev to the levels around 4.9 ev.

(c)
8.9 ev

(c) With bombarding energy raised to 8.9 ev, two more lines of intermediate wavelength appear. The photons have energies given by the transitions from the levels around 8.8 ev to the levels around 4.9 ev.

(d) As the electron-accelerating voltage is further increased, more and more lines appear. At 10.4 ev the atom is ionized and the complete spectrum is produced. In this case the classifications above and below the photo help to sort out the various sets of lines.

(d)

9.9 ev

10.4 ev

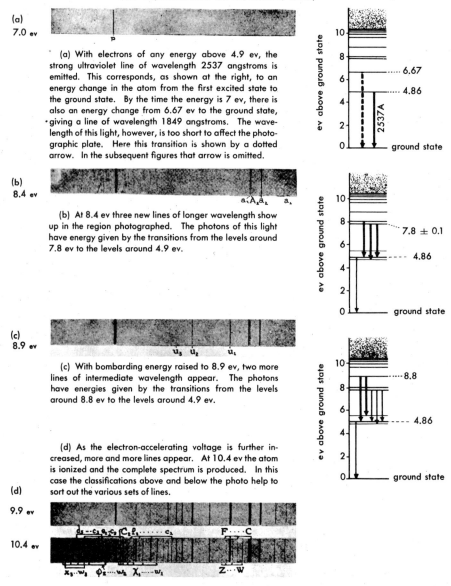

Fig. 15–17 The spectrum of mercury

of light—not all values of energy are acceptable to the atoms. From the results of Experiment 15–11 it is possible to determine the energy values which are acceptable to the atoms of a particular gas (Fig. 15–17). The energy values are in fact those which can be accepted by the orbital electrons. On accepting energy an electron moves from the ground state to the first higher-energy level—the atom is said to be **excited.** After a short time at this level the electron falls back to the ground state and emits the energy as a photon, a particle of light. If sufficient energy is given to the atom, an electron may rise to the second higher-energy level and then proceed to fall back to the ground state by emitting a photon of the appropriate energy. It could, however, fall back to the ground state in two steps stopping at the first energy level *en route*. If sufficient energy is supplied to the atom, the electron leaves it completely—this is known as **ionization.** When this occurs in the experiment, there is a large increase in current due to the increase in the number of free electrons.

(Experiment 15–12.) The **thyratron** is used as a switching device since there is a sudden and large increase in current at a particular anode potential. The large increase occurs when ionization is produced in the valve. Similarly a gas-filled photocell can give much larger currents than a vacuum photocell if the anode potential is above the ionization potential of the gas used.

Summary

Electrons can be emitted by solids under the action of heat, light, electron impact, or intense electric fields. In the diode electrons are accelerated to the anode by the anode having a positive potential with respect to the cathode. A current will flow only from cathode to anode and thus a diode can be used to rectify an alternating current.

The force experienced by a particle with a charge q in an electric field of intensity E is Eq. In a magnetic field of flux density B at right angles to the electron velocity the force on the electron is Bqv, where v is the velocity of the electron; the force due to the magnetic field, the magnetic field direction, and the electron velocity are mutually at right angles. The velocity of an electron accelerated from rest through a potential difference V is $(2Vq/m)^{1/2}$, m being the mass of the electron. The deflection of an electron beam in a magnetic field can be used to give a value of the charge-to-mass ratio for the electron. The motion of a charged oil drop in an electric field can give a value for the charge on the electron, $1\cdot6 \times 10^{-19}$ C, and hence a value for

Experiment 15–12

The thyratron is a gas-filled valve. Assemble the circuit (Fig. 15–18) so that a gradually increasing potential can be applied between the cathode and anode to accelerate the electrons. Measure the anode current. Plot a graph of current against the accelerating potential. Why is the graph different from that for the vacuum diode (Experiment 15–2)?

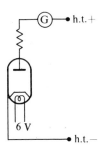

Fig. 15–18 Thyratron circuit (this is a simplified diagram of a thyratron omitting the parts not used)

the mass of the electron, 9.1×10^{-31} kg. The deflection of electron beams by electric and magnetic fields is used in **electron microscopes** and **cathode-ray oscilloscopes.**

An atom is said to be **excited** when one of its orbital electrons has accepted energy. The electrons in an atom can only accept particular values of energy; the accepted energy can be re-emitted as a particle of light, a **photon.** An atom is said to be **ionized** when it has been excited so much that it has lost an electron—in this state it will have a net positive charge. In photoelectricity and in the emission of light we are concerned with light as particles and not as waves. The energy of a photo-electron is directly proportional to the frequency of the incident light and independent of the intensity of illumination. The constant of proportionality, **Planck's constant,** is $h = 6.62 \times 10^{-34}$ J s.

Problems

15–1 How much kinetic energy is acquired by an electron when it is accelerated through a potential difference of 1 kV? $e = 1.6 \times 10^{-19}$ C.

15–2 Electrons are accelerated through a potential difference and then pass into a magnetic field. What path will the electrons follow when the magnetic field is (a) along the direction of the electron beam, (b) at right angles to the beam, and (c) at an angle of 45° to the beam?

15–3 What magnetic flux density is necessary to bend a beam of electrons, accelerated through 3 kV, into a circle of radius 19 cm? $m = 9.1 \times 10^{-31}$ kg; $e = 1.6 \times 10^{-19}$ C.

15–4 What electric field strength is necessary to balance the force of gravity and hold a free electron stationary in space? $m = 9.1 \times 10^{-31}$ kg; $e = 1.6 \times 10^{-19}$ C; $g = 9.8$ m/s.

15–5 What happens when electrons of energy (a) 0.5 eV, (b) 1.6 eV, and (c) 2.3 eV are incident on a surface having a work function of 1.6 eV?

15–6 Between what limits will be the velocities of the emitted electrons when light of wavelength 5×10^{-1} m is incident on a surface of work function 1.8 eV? $h = 6.62 \times 10^{-34}$ J s.

15–7 The minimum rate of incidence of radiation of wavelength 5.5×10^{-7} m that can be detected by the average eye is 2×10^{-16} W. What is the minimum number of these photons hitting the eye per second to produce vision? $h = 6.62 \times 10^{-34}$ J s.

*15–8 What is a photon? Is it a particle? Is it a wave?

15–9 Explain what happens as a gradually increasing amount of energy is supplied to electrons used to bombard atoms.

15–10 What happens when electrons of energy (a) 7 eV, (b) 10.18 eV, and (c) 11.0 eV bombard hydrogen atoms? First energy level of hydrogen is 10.18 eV; second level is 12.07 eV.

15–11 How does an elastic collision differ from an inelastic collision?

15–12 Explain how the electron lens system in a cathode-ray oscilloscope produces a focused image on the fluorescent screen? How is the brightness of the image controlled?

15–13 Describe an arrangement with a photocell which can be used to count the number of objects passing a particular point on a conveyor belt.

15–14 How does the Millikan experiment show that charge comes in certain definite quantities, that is that charge is particulate?

16 Radioactivity

16-1 The emission of radiation

X-rays were originally produced when a beam of electrons struck the glass wall of an evacuated tube, and this rapid deceleration of the electrons gave rise to the emission of X-rays. Where the electrons struck the tube wall, a strong fluorescence was also observed. It seemed possible that the fluorescence and the emission of X-rays might be related and that where fluorescence occurred X-rays were also produced. In 1896 Becquerel examined a number of materials that fluoresced. His procedure was to wrap a photographic plate in thick black paper and place on top of the paper the appropriate fluorescent material. Roentgen had already discovered that X-rays were able to penetrate black paper and to blacken a photographic plate and thus Becquerel was hoping to obtain blackened photographic plates. The covered plate and the fluorescent material were placed in the Sun because fluorescence only occurred after exposure to the Sun or a strong light. After some negative results blackened plates were produced with uranium compounds. However, the uranium compounds blackened the plate regardless of whether they fluoresced or not—fluorescence and blackening of photographic plates were not related. In fact fluorescence and blackening of plates are not related in the case of X-rays. The property possessed by the uranium compounds which caused the photographic plates to blacken is called **radioactivity.**

(Experiments 16–1 and 16–2.)

In addition to their effect on a photographic plate it was soon found that the radiation from radioactive materials was able to ionize air. This gave a convenient method for detecting the radiation.

When a radioactive material such as uranium or thorium is placed near a charged object, leakage of the charge occurs owing to the air molecules becoming ionized by the radiations emitted by the material. A more convenient detector of the ionization is the ionization chamber (Fig. 16–1). In the simple form of this instrument there is a central metal rod insulated from an outer cylindrical metal electrode. These two electrodes are connected to a voltage supply of about 150 V, the central rod being positively charged. The ionization chamber is in fact a capacitor of very low capacitance. When an ionizing radiation passes through the chamber, a small current flows between the elec-

---

Experiment 16–1

Place either a photographic plate or a sheet of bromide paper in a light-proof envelope. On top of the envelope place a metal object such as a key. Spread over the object and a part of the plate a layer of a uranium or thorium compound and leave for about 4 days. At the end of this time develop the plate in the normal manner.

Describe the appearance of the plate and offer an explanation of the results.

Experiment 16–2

An electroscope should be given a charge so that the leaf is deflected. Introduce a uranium or thorium salt into the electroscope chamber. The material should be in a small open container and care should be taken not to spill any. What happens to the charged leaf when the source is introduced? Cover the uranium or thorium with a piece of paper. Is there any change in the behaviour of the charged leaf? (Note. A pulse electroscope could be used.)

---

trodes. This current can be detected by an electroscope with its leaf connected to the central rod and its casing connected to the outer casing of the ionization chamber. In practice a robust form of ionization chamber is used in which the leaf is a quartz fibre. This type is known as a **dosimeter** and is often carried in the pocket of a person working with ionizing radiations to check the quantity or dose of radiation they may have received. The rate at which the quartz fibre moves across an internal scale is proportional to the rate at which ions are produced in the chamber.

(Experiment 16–3.)

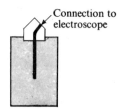

Fig. 16–1 Ionization chamber

When the uranium is covered with a piece of paper there is a marked reduction in the rate at which ions are produced. If two pieces of paper are used, there is very little difference from the result obtained with just one piece—ions are still produced but their numbers with one or two pieces of paper are considerably less than the results without paper. The conclusion that can be drawn from this is that the radiation emitted by uranium consists of at least two types—one that is intensely ionizing and is absorbed by a piece of paper and the rest which are not so ionizing and penetrate paper easily, very little absorption occurring. The intensely ionizing readily absorbed radiation is known as **alpha radiation.**

In the ionization chamber the ions produced by the radiation collide with numerous atoms on their way to the appropriate electrode and thus do not reach very high velocities. If the potential difference between the two electrodes is increased, the chamber made gas tight, and the internal pressure reduced, the ions can reach much higher velocities. This can result in the ions having sufficient energy to cause ionization of gas molecules by collision. If this process is repeated many times, an **avalanche** of charged particles is produced. A tube in which this occurs is known as a **Geiger tube** (Fig. 16–2).

(Experiment 16–4.)

When the applied voltage exceeds a certain value, known as the **threshold,** the pulses produced by the incidence of radiation on the tube become large, of the

---

Experiment 16–3

A dosimeter form of ionization chamber should be used for this experiment. The instrument should be connected to a d.c. voltage supply. Slowly increase the applied voltage and with the switch to the apparatus closed observe the quartz fibre. At a voltage of about 150 V the fibre will move across the scale—move it to the zero position. When this is accomplished, the ionization chamber will be fully charged, and the switch should be opened. Observe the fibre and note the rate at which it moves across the scale. Very little motion should occur. If discharging occurs, the inside of the chamber should be cleaned with alcohol as

leakage of charge is probably occurring across the insulation.

Introduce a uranium compound. What happens to the quartz fibre? Cover the uranium with a piece of paper; what change occurs? Try a second piece of paper. Cover the uranium with a piece of aluminium, about 1 mm thick. What happens? Is there any difference in the results with the paper and with the aluminium?

Experiment 16–4

Assemble the circuit as indicated in the diagram (Fig. 16–3). A d.c. potential difference is applied to the Geiger tube. The incidence of radiation on the tube and

order of volts, owing to the avalanche effect. If these pulses are applied to a counting circuit, the threshold corresponds to the voltage at which counting starts, generally for pulse input voltages of about 0·2 V. This circuit is the basis of an instrument known as a **scaler.**

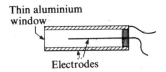

Fig. 16–2 Geiger tube

The detected pulse rate, termed the **count rate,** for a particular source-to-tube distance increases as the voltage is increased above the threshold, until a plateau is reached where the count rate is almost independent of the applied voltage—this is the region in which the tube is operated (Fig. 16–4). If the voltage is still further increased, the tube goes into continuous discharge and may be damaged. A Geiger tube should always be operated in the middle of the plateau, that is about 50 V above the threshold.

(Experiment 16–5.)

If instead of counting the pulses they are fed to a capacitor and the leakage current through a resistance measured, the instrument is known as a **ratemeter,** the

current being related to the count rate. The time over which the ratemeter averages the count is known as the **integration time** and is determined by the time constant of the circuit, that is the product of the capacitance and the resistance.

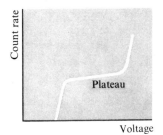

Fig. 16–4 Geiger tube characteristic

(Experiment 16–6.) A radioactive source such as uranium or radium emits three different types of radiation—**alpha, beta,** and **gamma.** Alpha radiation is stopped by a thin sheet of paper and will only have been detected by the usual Geiger tube if the source is placed very close to the tube window. Beta radiation is more penetrating and is only stopped by about 1 mm of aluminium. Gamma radiation is very penetrating and requires a sheet of lead to stop it. Of the three radiations

---

its resulting effect are detected by either an oscilloscope or a pair of headphones placed across the tube. A capacitor is placed between the tube and the headphones to block the high d.c. voltage.

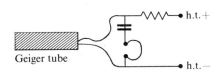

Fig. 16–3 Circuit for Experiment 16–4

Observe the result when a radioactive source is placed near the tube, and the applied voltage is gradually increased from 300 to 450 V.

Experiment 16–5
Connect a Geiger tube to a scalar and place a radioactive source a fixed distance from the tube window. Determine how the count rate varies with the applied voltage. Plot a graph of count rate against voltage and hence obtain the voltage of the threshold and the middle of the plateau.

alpha is the most ionizing and gamma the least. Because of the high penetrating and low ionizing ability of gamma radiation the Geiger tube is a very inefficient detector for gamma radiation.

(Experiment 16–7.) By measurement of the variation of count rate of the transmitted radiation with thickness of absorber, automatic gauges can be devised for the measurement of thickness without the sample being touched (Fig. 16–5). With alpha sources the absorber must be very thin, with beta foils up to about 1 mm can be used, while with gamma thick plate can be measured.

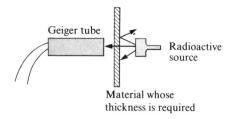

Fig. 16–5 Thickness gauge

Another radiation detector is the **cloud chamber.** Water droplets form most easily on dust particles and charged particles such as ions. In the absence of such condensation centres supersaturation will occur before any cloud of water drops will form. In the continuous cloud chamber an enclosure is cooled to a low temperature by solid carbon dioxide (Figs. 16–6 and 16–7).

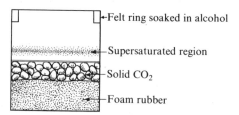

Fig. 16–6 Continuous cloud chamber

Alcohol vapour is introduced into the chamber from a felt ring soaked in alcohol. Because of the low temperature supersaturation occurs. Ions or dust particles in the chamber will cause condensation and thus reveal their presence. When ionizing radiation passes through the chamber, its path shows up as a line of alcohol drops due to alcohol condensing on the ionized molecules along the path of the radiation.

(Experiment 16–8.)

16–2 Identification of the radiation
Electromagnetic waves are not deflected by either electric

Experiment 16–6
Place a uranium or radium source close to the Geiger tube window. Determine the count rate with either a scalar or a ratemeter, the Geiger tube being operated in the middle of the plateau. Place a piece of paper between the source and the tube and again determine the count rate. Replace the paper with a sheet of aluminium about 1 mm thick and determine the count rate. Repeat the measurement with a sheet of lead in place of the aluminium. Use more lead sheets. Comment on the results.

Experiment 16–7
With the aid of a beta-emitting source, ^{90}Sr, and a Geiger tube with scaler or ratemeter construct a gauge which can be used to determine the thickness of aluminium sheet.

Fig. 16-7

or magnetic fields. It is, however, possible to deflect particles such as the electron because of its charge—uncharged particles would not be deflected. Thus the deflection of a radiation in either a magnetic or an electric field would be evidence that the radiation consisted of charged particles.

(Experiment 16–9.) Alpha and beta radiation are both deflected in magnetic and electric fields—thus indicating their charged nature. Gamma radiation is not deflected in either magnetic or electric fields. Because alpha and beta radiations are charged, their ratio of charge to mass can be found. Beta radiation is found to have the same value of q/m as occurs with fast electrons emitted by thermionic emission and to have a charge of the same sign. Alpha radiation is found to be positively charged particles with a value of q/m which is the same as that of a helium atom which has lost both its electrons. This identification was confirmed when the alpha radiation was found to give the same spectrum as helium.

Gamma radiation is short-wavelength X-radiation but, because of its higher energy, is more penetrating than the longer waves.

16–3 Radioactive decay

(Experiment 16–10.) Because of the randomness of radioactivity, as regards both direction of emission and the time at which emission occurs, the count recorded in any one interval of time is unlikely to be the same in the next interval of time. The scatter of results can be represented by the standard deviation. This is the square root of the recorded count, e.g. a count of 100 has a standard deviation of 10. This means that there is a 68% possibility that the count when repeated will lie within 100 ± 10. There is a 95.4% possibility that the

---

Experiment 16–8

Put solid carbon dioxide in the bottom of the cloud chamber and wet the felt ring with alcohol. To remove ions and dust from the chamber rub the Perspex lid with a cloth—the lid then becomes electrostatically charged and attracts dust and ions. After a few minutes tracks should be seen emanating from the radium source supplied with the chamber. If no tracks are seen, rub the lid again.

Radium decays to give alpha, beta, and gamma radiations—however, the main tracks visible will be alpha tracks due to the large number of ions produced by such radiation. How much air can the alpha radiation penetrate? Do the tracks show kinks where the alpha radiation encountered air or alcohol molecules?

Experiment 16–9

Direct a beam of beta radiation through a hole in a lead plate onto an end window Geiger tube (Fig. 16–8).

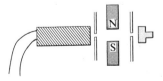

Fig. 16–8 Investigation of the charged nature of radiations

count will lie within two standard deviations, that is 100 ± 20. These results can be confirmed by an examination of the results of Experiment 16–10. Because of this randomness, counts as large as possible should be taken in order to increase the percentage accuracy of the count rate. For example, see the following.

Count	Standard deviation	Percentage accuracy (%)
100	10	...10
10,000	100	...1
1,000,000	1,000	...0·1

Thorium compounds emit a radioactive gas generally known as **thoron.** This gas emits alpha radiation. A convenient generator of this gas can be made by placing thorium carbonate or thorium hydroxide in a chemistry wash bottle (Fig. 16–9). When the plastic bottle is gently squeezed, the radioactive gas is ejected—a small piece of cloth placed in the bottle prevents the thorium compound being ejected.
(Experiment 16–11.)

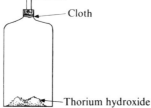

Cloth

Fig. 16–9   Thoron generator — Thorium hydroxide

When a radioactive material emits radiation, it does so regardless of the chemical or physical state of the material, that is the radiation emitted, for example, by uranium is the same whether the uranium is in the form of a metal or a compound. The radiation depends only on the element concerned (in fact the particular isotope concerned, see later for an explanation). When a particular atom has emitted its radiation, it changes to another type of atom: we say the atom has **decayed.** The rate at which the decay occurs depends on the element (isotope) concerned and the amount present. Let N be the number of atoms of a particular radioactive substance present at some instant of time.

$$\text{rate of decay} = \frac{dN}{dt}$$

Experimental evidence shows that

$$\frac{dN}{dt} \propto -N$$

Hence inserting a constant λ which is characteristic of the radioactive substance concerned (known as the **decay constant**)

$$\frac{dN}{dt} = -\lambda N$$

The minus sign is included because we are concerned

Bring a magnet near the radiation beam—is there any change in count rate indicating deflection of the radiation? Try gamma radiation.

Try the same experiment with an alpha source and a solid-state detector.

Experiment 16–10
Place a radioactive source close to a Geiger tube window and with the aid of a scalar determine the count in each of a number of successive 1-min intervals. Are the results all the same or is there some obvious trend? Is radioactivity a random process?

Plot a graph showing the distribution of the count, that is a graph of the number times a particular count occurs against the count.

with a decay, things getting smaller. dN is the number by which N is reduced in time dt. dN/dt is the rate of disintegration and is known as the **activity.** The unit of activity is the **curie.** One curie (Ci) is a disintegration rate of 3.7×10^{10} atoms/s. This is a high activity and millicuries (10^{-3} Ci) and microcuries (10^{-6} Ci) are more general in industrial applications. Rearranging the equation,

$$\frac{dN}{N} = -\lambda \, dt$$

and thus integrating between the limits of N_0 atoms at zero time and N at time t

$$\int_N^{N_0} \frac{dN}{N} = \int_0^t -\lambda \, dt$$

Therefore

$$\ln\left(\frac{N}{N_0}\right) = -\lambda t$$

Hence

$$N = N_0 \, e^{-\lambda t}$$

This relationship shows that the number of radioactive atoms of a particular substance varies with time (Fig. 16–10) in an exponential manner.

(Experiment 16–12.) The **half-life** of a radioactive substance is the time taken for the number of active atoms to decrease by half. In the half-life T the number of atoms changes from N_0 to $\frac{1}{2}N$.

$$\tfrac{1}{2}N_0 = N_0 e^{-\lambda T}$$

Hence

$$\tfrac{1}{2} = e^{-\lambda T} \qquad \text{or} \qquad \ln\left(\tfrac{1}{2}\right) = -\lambda T$$

and thus

$$T = \frac{\ln(2)}{\lambda} = \frac{0.693}{\lambda}$$

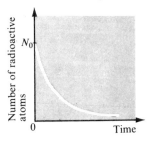

Fig. 16–10 *Variation of the number of radioactive atoms, in a sample, with time (decay of radioactivity)*

Experiment 16–11

Allow a small quantity of thoron to enter the cloud chamber—alpha tracks should be visible. How does the number of alpha tracks vary with time (very roughly)? Is something more than randomness of emission necessary to explain the results?

Experiment 16–12

Take a large number of dice, about a hundred, shake, and then throw them. Reject all dice that show a 6 and consider those dice to have decayed. Repeat the process with the remaining dice. Plot a graph showing how the number of 'active' dice varies with the number of throws. How does the activity, that is rate of decay, vary with time? What is the time, that is number of throws, necessary for the number of dice to decrease by half? Does this result depend on the initial number of dice?

Example 16–1. How many disintegrations occur per second for a 5-μCi source?

1 Ci is $3 \cdot 7 \times 10^{10}$ disintegrations/s. Hence 5 μCi is $5 \times 10^{-6} \times 3 \cdot 7 \times 10^{10}$ disintegrations/s, that is $18 \cdot 5 \times 10^4$ disintegrations/s.

Example 16–2. A 5-μCi source has a half-life of 6 days. How long must elapse before the activity is (a) $2 \cdot 5$ μCi, and (b) $1 \cdot 25$ μCi?

In 6 days the activity decreases by half. Therefore 6 days is the time that must elapse for the activity to decrease to $2 \cdot 5$ μCi.

A further 6 days will be necessary for the activity to change from $2 \cdot 5$ to $1 \cdot 25$ μCi. Thus 12 days is the time taken for the activity to fall from 5 to $1 \cdot 25$ μCi.

As the activity is directly proportional to the number of atoms, then the time taken for the activity to decrease by half will be the same as that necessary for the number of atoms to decrease by half.

Example 16–3. The carbon isotope ^{14}C decays by beta emission with a half-life of 5,700 years. The carbon is continually being produced in the Earth's atmosphere and is taken in by living organisms. As soon as an organism dies, it ceases taking in fresh carbon and the carbon already present decays away. Thus the amount of ^{14}C present in a dead organism can be used to date it. Determine the age of a sample which gives $0 \cdot 04$ counts per minute per gramme of fresh carbon compared with 15 counts per minute per gramme of fresh carbon.

For fresh carbon

$$15 = \frac{dN_1}{dt} = -\lambda N_1$$

N_1 is the number of active atoms. For the old sample

$$0 \cdot 04 = \frac{dN_2}{dt} = -\lambda N_2$$

N_2 is the number of active atoms remaining in the old sample but

$$N_2 = N_1 e^{-\lambda t}$$

therefore

$$0 \cdot 04 = 15 e^{-\lambda t}$$

$$\ln(0 \cdot 04) = \ln(15) - \lambda t \qquad \text{but} \qquad \lambda = \frac{\ln(2)}{T}$$

Hence

$$\lambda = \frac{\ln(2)}{5,700 \times 365 \times 24 \times 60} \ \text{s}^{-1}$$

Experiment 16–13
Devise an instrument which could be used to measure the depth of water in a container. Try it out.

Experiment 16–14
Investigate the efficiences of detergents in removing grease from cloth. Uranium oxide or thorium carbonate (insoluble in water) can be used as the tracer.

and therefore

$$t = \frac{\ln(15) - \ln(0 \cdot 04)}{\ln(2)/(5{,}700 \times 365 \times 24 \times 60)}$$

Hence t is $2 \cdot 33 \times 10^{10}$ min or $4 \cdot 44 \times 10^4$ years.

16–4 Radioactive transformations

When thorium decays, a new element radium is formed; the decay is by means of alpha particles:

$$^{232}_{90}\text{Th} \rightarrow {}^{228}_{88}\text{Ra} + {}^{4}_{2}\text{He}$$

The above equation describes the process. The lower number to the left of the element symbol is the amount of positive charge held by an atom of the element, one unit of charge being equal in magnitude to the fundamental charge, that is the charge on the electron. The number of positive charges in an atom is known as the **atomic number.** The upper left-hand number is the mass of an atom expressed on the atomic mass unit scale, the mass of a hydrogen atom being approximately one.

As charge is conserved, the lower line of the equation must balance and, if mass is conserved, the upper line must also balance. An alpha particle is an ionized helium atom with two units of positive charge and four units of mass—hence ${}^{4}_{2}\text{He}$.

Radium is also radioactive and emits beta radiation:

$$^{228}_{88}\text{Ra} \rightarrow {}^{228}_{89}\text{Ac} + {}_{-1}\text{e}$$

The charge on the electron is -1 and its mass is negligible by comparison with masses involved in the equation. Ac is the element actinium. This is also radioactive and decays by beta emission to give thorium:

$$^{228}_{89}\text{Ac} \rightarrow {}^{228}_{90}\text{Th} + {}_{-1}\text{e}$$

The identity of an element is determined by the number of positive charges held by its atoms. Thorium always has 90. There is, however, a difference in the masses of the thorium produced in this decay and the thorium which started this series of decays. One has a mass of 228 and the other a mass of 232. Chemically these two

forms of thorium are indistinguishable and are known as **isotopes.** There are other differences between the two thorium isotopes; ${}^{228}\text{Th}$ has a half-life of $1 \cdot 90$ years and ${}^{232}\text{Th}$ $1 \cdot 39 \times 10^{10}$ years.

Radioactive and new stable isotopes of most elements can be produced by the bombardment of stable elements. For example, when nitrogen is bombarded by alpha particles, an isotope of oxygen is produced:

$$^{14}_{7}\text{N} + {}^{4}_{2}\text{He} \rightarrow {}^{17}_{8}\text{O} + {}^{1}_{1}\text{H}$$

This oxygen isotope is stable though not the usual form of oxygen. The single positive charge ${}^{1}_{1}\text{H}$ is a proton (a hydrogen atom without its electron). Bombardment of beryllium with alpha particles produces carbon and an uncharged radiation with a mass very close to that of a proton:

$$^{9}_{4}\text{Be} + {}^{4}_{2}\text{He} \rightarrow {}^{12}_{6}\text{C} + {}^{1}_{0}\text{n}$$

The uncharged particle is a **neutron.**

${}^{27}\text{Al}$ when bombarded by neutrons produces radioactive sodium which decays with a half-life of $14 \cdot 8$ days to produce magnesium:

$$^{27}_{13}\text{Al} + {}^{1}_{0}\text{n} \rightarrow {}^{24}_{11}\text{Na} + {}^{4}_{2}\text{He}$$

$$^{24}_{11}\text{Na} \rightarrow {}^{24}_{12}\text{Mg} + {}_{-1}\text{e}$$

16–5 Uses of radioactive isotopes

There are numerous uses of radioactive isotopes and only an indication of these uses can be given here.

The measurement of thickness, densities, and levels
These depend on the variation in count rate for radiation transmitted through, or reflected from, a material. The count rate depends on the thickness and the density of the material (see Experiment 16–7). Beta sources are widely used, e.g. in the thickness control of paper, plastics, thin metal sheets, etc. Gamma sources are used where large thicknesses of matter are concerned, e.g. the checking of the homogeneity of metal ingots.

Levels of liquids can be measured by passing the radiation vertically through the liquid and measuring the change in absorption due to the varying depth. Another possibility is to fix a source to a float and to place a detector at the top of the vessel, as the float moves upwards so the count rate increases.

Radioisotopes as tracers
It is possible to detect very small amounts of radio-active materials—a single beta particle can be detected and this would be from a single atom. The general procedure is to 'label' an item with an appropriate isotope; this item can then be followed through different processes. Typical examples of this would be a measurement of the production of carbon dioxide in a car engine by labelling the fuel with ^{14}C, the effectiveness of a washing process by using a labelled dirt or grease, the detection of leaks from an underground pipe by passing a radioactive fluid along the pipe, the study of the efficiency of mixing of the constituents of an alloy, measurement of wear with machine tools having been made active by neutron bombardment, etc.

Gammaradiography
The same technique as that for X-rays can be used for detection of the radiation. The radiation is passed through the sample onto a photographic film—and differences in internal structure such as flaws show up as a dark patch on the film.

(Experiments 16–13 and 16–14.)

16–6 Radiation hazards
The human body is daily subject to irradiation from natural sources and thus the hazards involved in using radioactive isotopes in experiments must be compared with that from the natural level. The main effects of radiation are the production of ions in the body which disturb the existing equilibrium and, for example, change the speed of reproduction of cells, or which cause the destruction of cells, surface damage similar to a burn or a skin cancer, and genetic effects which may show as abnormal hereditary characteristics. Fortunately it is possible to do a lot of radioisotope work with sources of very low activity and also to use shielding.

Alpha radiation requires almost no shielding as a sheet of paper or the skin is capable of stopping it. Beta radiation requires generally about a millimetre of aluminium, while gamma radiation is difficult to shield against without large thicknesses of lead. The above refers to the radiation falling on the external surface of the body—if the radiation is from an ingested source, the problem is different. Internal irradiation is the most dangerous and should not be allowed to occur in the laboratory. Alpha radiation in producing the most ions per unit length of path is the worst; beta sources are also a significant hazard when taken internally. All sources internally are dangerous as there is no shielding and the sources can be very close to vital organs. One point to bear in mind with external sources is that the dose received by the body varies roughly as the inverse square of the distance from source to the body. Therefore the greater the distance the smaller is the dose received.

The absorbed dose unit is the **rad,** one rad being defined as the absorption of 10^{-5} J of energy by one gramme of substance. The maximum acceptable dose rate for normal laboratory work in school or college laboratories is 50 mrad/year for any one individual. With the sources generally available the following are the doses per year at a distance of 10 cm from the source:

^{60}Co	$5\mu Ci$	6·0 rad/year
^{226}Ra	$5\mu Ci$	3·7 rad/year
^{90}Sr	$5\mu Ci$	11·7 rad/year

Thus for ^{60}Co the maximum permissible dose for a year can be received by exposure to the source at a distance of 10 cm for about 3 days in that year. At a distance of 5 cm this dose is received in about 18 hours. The obvious deduction from this is that the exposure times should be kept as small as possible and the distance of the source from the body as large as possible.

Summary

Natural radioactive sources decay by the emission of **alpha particles** (helium nuclei), **beta particles** (electrons), or **gamma rays** (electromagnetic waves of short wavelength). The radiations can be detected by their effects on photographic emulsions, cloud chambers, or detectors using the ion current produced by the radiation, e.g. ionization chamber, Geiger tube.

The radiations are emitted from a source in a random manner and this must be taken into account when considering the accuracy of any result. Radioactive isotopes decay according to an exponential relationship:

$$N = N_0 e^{-\lambda t}$$

The **half-life** is the time taken for the number of radioactive atoms to decrease by half:

$$\text{half-life} = \frac{\ln(2)}{\lambda}$$

The **activity** of a sample is the number of disintegrations per unit time and is measured in curies:

$$\text{activity} = \frac{dN}{dt}$$

One curie is 3.7×10^{10} disintegrations/s.

As a result of the emission of radiation a change in the source occurs. When an alpha particle is emitted, the atomic number of the atom decreases by two and the atomic mass by four. When beta emission occurs, the atomic number increases by one and there is no significant change in the atomic mass. No change in either atomic mass or number occurs when gamma radiation is emitted.

Isotopes are elements with the same atomic number but a different atomic mass. Radioactive isotopes can be produced by the bombardment of stable elements.

Problems

16–1 The following information is generally present on the manufacturer's data sheet accompanying a Geiger tube. What is the significance of the information?

Threshold voltage	370 V
Plateau length	100 V
Plateau slope	7/100 %/V
Dead time	125 μs
Window thickness	3·5 to 4·0 mg/cm^2

16–2 How many disintegrations occur per second for a 5-μCi source?

16–3 Radon (atomic number 86, atomic mass 222) has a half-life of 3·8 days.
   (a) How much of the initial mass is present after 3·8 days?
   (b) How much of the initial mass is present after 7·6 days?
   (c) How much time must elapse before all the radon has disintegrated?
   (d) What is the atomic number and mass of the new element produced if the decay of radon is by alpha emission?

16–4 The activity of a source of strontium 90 is 5 mCi and the half-life is 20 years. What is the activity after 80 years?

16–5 0·012% of natural potassium is a radioactive isotope. It has a half-life of 1.27×10^9 years, 11% decaying to argon and the other 89% to calcium. Nearly all the terrestrial argon is considered to have been produced by the decay of potassium. The mass of argon present in the Earth's atmosphere is 1·3% of the total mass of the atmosphere and the estimated mass of potassium in the Earth is 0·026% of the mass of the Earth. Estimate the age of the Earth. Mass of the atmosphere $= 5.27 \times 10^{18}$ kg; mass of the Earth $= 5.98 \times 10^{24}$ kg.

16–6 The following count-rate results were taken for a decaying element.

Time (s)	1	100	200	300	400	500
Count rate (s^{-1})	2,000	1,638	1,340	1,098	898	736

What is the half-life of the element?

16–7 How could you distinguish in the laboratory between alpha, beta, and gamma radiations?

16–8 Explain the apparatus and the arrangement necessary for the following measurements: (a) a thickness gauge for the determination of the thickness of aluminium foil, (b) an investigation into the efficiency with which cement and sand are mixed in the production of concrete, (c) a gauge for the determination of liquid levels in steel containers, and (d) flaw detection in sheet steel.

16–9 Natural potassium contains a small percentage of a radioactive isotope. This emits beta radiation of relatively low energy. Explain how you would detect the activity of a potassium sample.

16–10 A ^{60}Co source used in a thickness gauge has a half-life of 5·3 years. If the initial activity of the source is 4 mCi, how long must elapse before the activity falls to 0·5 mCi?

17 The Atom

17–1 Crystals

(Experiment 17–1.) Most solids are crystalline—these include the obvious crystals such as common salt and sugar, together with the metals. Most metal samples are a mixture of a large number of small crystals, that is a polycrystalline structure similar to the structure of the solidified mass of hypo but generally with much smaller crystals. A characteristic of crystals is that corresponding faces of different crystals of the same substance always make the same angle with each other. For example the angles between faces of a common salt crystal are always 90°.

(Experiment 17–2.) At certain particular angles many crystals cleave, that is they break easily along a plane and leave smooth surfaces. It is as if they were like a book which opens easily at a page but is not so easily opened at right angles to a page—tearing has to occur. This, together with the regular structure of crystals, suggests that perhaps solids are put together from building blocks in a regular manner.

(Experiments 17–3 and 17–4.) The building blocks for crystals are atoms. Information regarding their arrangement can be obtained from X-ray crystal-lography. When X-rays are reflected or transmitted by crystals, an interference pattern is produced. The positions of constructive interference are determined by the spacing of the atoms in much the same way as the spacing of the rulings on a diffraction grating determine the constructive interference positions for light. (For revision see Experiment 12–2 and Section 12–6.) The results show that the spacing of atoms is of the order of 10^{-10} m in solids.

Example 17–1. Consider a cubic crystal in which the atoms are arranged at the corners of a cube of side 2×10^{-10} m. What X-ray diffraction results would you obtain from the different atomic planes with X-rays of wavelength 0.5×10^{-10} m?

Thus

$$2d \sin \theta = n\lambda$$

gives for the 2×10^{-10} m spacing

$$\sin \theta = 0.125, 0.250, 0.375, \text{ etc.}$$

These values correspond to the orders, $n = 1, 2,$ and 3 respectively.

Experiment 17–1

Examine various crystals under a microscope or a magnifying glass. What similarities are there between crystals of the same material? Are there similarities between different-sized crystals of the same material? Possible materials for examination are common salt, sodium thiosulphate (hypo), sugar, etc.

Melt a small amount of sodium thiosulphate on a microscope slide by gentle heating. Observe the liquid while it is cooling and crystals reform. How does a crystal grow? Does the final mass of crystals appear crystalline to the naked eye and also under the microscope? Are metals crystalline? Examine the surfaces of metals—side illumination of the surface is necessary. Melt a piece of lead and allow it to cool. Do crystals form?

Experiment 17–2

Set a large crystal of calcite in a piece of Plasticine to keep it stationary. Then place a single-edge razor blade at various angles on the surface, and apply a sharp tap to the blade. What happens?

Experiment 17–3

Stack polystyrene spheres in a regular manner so that a stable structure results. Could this be the way that crystals are built up?

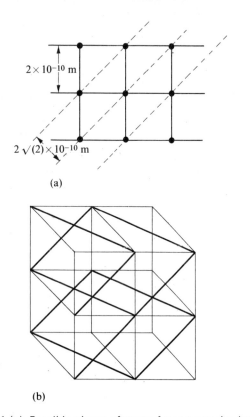

(a)

(b)

Fig. 17–1 (a) *Possible planes of atoms for constructive interference reflections.* (b) *Another possible set of planes*

Quite a large number of possible planes exist between which interference effects can occur even in something as simple as a cubic crystal (Fig. 17–1). There are, for example, planes with atoms 2×10^{-10} m apart and $\sqrt{(2)} \times 10^{-10}$ m apart. For the $\sqrt{(2)} \times 10^{-10}$ m spacing

$$\sin \theta = 0 \cdot 176, \ 0 \cdot 352, \ 0 \cdot 528, \text{ etc.}$$

The strength of the reflected signal depends on the number of atoms in the layers concerned and the order number. The greater the number of atoms involved, the greater is the reflected signal strength; the lower the order number, the greater is the signal strength. Thus the 2×10^{-10} m results will be more intense than the $\sqrt{(2)} \times 10^{-10}$ m results for the same order. (Experiment 17–5.)

17–2 Constituents of atoms

When an acetate strip is rubbed with a cloth, a transfer of charge occurs and the charged nature of each can be detected by their ability to attract small pieces of paper. That the charge on the cloth is opposite and equal to that on the strip can be shown with an electroscope. The obvious conclusion to draw is that the atoms are initially neutral owing to their having equal amounts of positive and negative charge, friction having removed some of the more loosely held negative charges.

Experiment 17–4
Blow bubbles with a fine tube in a soap solution. The bubbles will be attracted to each other and form up in regular structures. Compare the pattern with that resulting from the solidification of hypo.

Experiment 17–5
This is a simple analogue experiment using microwaves in place of X-rays. Direct a beam of parallel radiation onto the 'polystyrene crystal' and focus the reflected radiation onto the detector by means of a paraffin wax lens. The 'polystyrene crystal' is constructed of an orderly stacked pile of polystyrene spheres. Determine

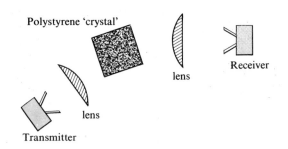

Fig. 17–2 *Microwave spectrometer for the determination of 'crystal' structure*

The nature of the charge has been found from experiments involving the deflection of the charges in electric and magnetic fields. These show that the charges that are removed are negative and have a charge of $1\cdot6 \times 10^{-19}$ C; they are known as **electrons** (Chapter 6). Thus, when an acetate strip is rubbed with a piece of cloth, electrons are transferred from one material to the other, leaving the first positively charged.

The mass of the electron is very small, $9\cdot1 \times 10^{-31}$ kg; what of the mass of the rest of the atom—the part carrying the positive charge? Deflection experiments can be done with instruments known as mass spectrometers in which electric and magnetic fields act on **ions,** atoms which have lost electrons. In a simple instrument there is an ion source where electrons are removed from atoms by bombardment with other electrons, an acceleration region where the positive ions are accelerated, and a region where the ion beam undergoes dispersion in a magnetic field (Fig. 17–3). Velocity selection is by means of an electric field and a magnetic field producing opposite forces on the ions, placed between the acceleration and dispersion regions.

force due to the magnetic field $= Bqv$

where B is the magnetic flux density, q the charge carried by the ion, and v the velocity of the ion.

force due to the electric field $= Eq$

where E is the electric field strength. The forces exerted by the electric and magnetic fields will be equal for one velocity only, given by

$$Bqv = Eq \qquad \text{when} \qquad v = E/B \qquad (17\text{–}1)$$

The force on the ions due to the magnetic field is Bqv. Hence

$$Bqv = mv^2/r \qquad (17\text{–}2)$$

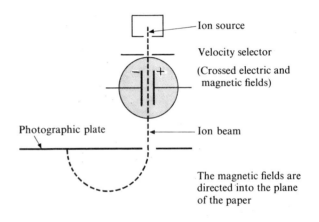

Ion source

Velocity selector

(Crossed electric and magnetic fields)

Photographic plate

Ion beam

The magnetic fields are directed into the plane of the paper

Fig. 17–3 Mass spectrometer

the angles at which reflection occurs for a particular face of the crystal (Fig. 17–2). Repeat the measurement for the other faces of the crystal and hence obtain the spacing between the 'atoms' in the crystal.

where m is the mass of the ion, and r is the radius of the path. Thus eliminating v from Eqns (17–1) and (17–2) gives

$$m = B^2 qr/E$$

For a given charge q the distance along the detector plate is proportional to the mass of the ion, and hence the scale is linear. The results of such measurements show that the mass of an atom is considerably greater than that of the electron. In addition, the results show the presence of **isotopes,** atoms of the same element and having the same chemical properties but different masses. The table gives percentages for the isotopes of oxygen.

Mass (a.m.u.)	Percentage of isotope (%)
16	99·76
17	0·037
18	0·20

1 a.m.u. (atomic mass unit) $= 1·66 \times 10^{-27}$ kg

Precise measurements show that the mass figures for elements are not exactly whole numbers but are quite close to them. The mass of an oxygen atom is almost thirty thousand times more than that of an electron. The major part of the mass of an atom is thus not provided by the electrons.

How many electrons are there per atom? Elements when arranged in a sequence dictated by their chemical properties, the **periodic table,** or by their X-ray spectra would appear to have an extra electron for each successive step taken along the sequence. Thus oxygen, number 8 in the table, has eight electrons while nitrogen, number 7, has seven electrons. As a consequence of this, oxygen has eight units of positive charge, while nitrogen has seven units. That atoms have more than one electron can be shown by applying a gradually increasing amount of energy—when a high enough level has been reached **ionization** occurs, that is an electron is ejected and so on. The number of positive charges carried by an atom is known as the **atomic number** for that atom.

Not all the mass in the atom is carried by the positive charges, known as **protons.** Uncharged particles also exist as shown by experiments involving the bombardment of atoms with alpha particles (Section 16–4). These neutral particles, known as **neutrons,** have a mass comparable with that of the proton. The mass of the proton can be determined by deflection methods such as in the mass spectrometer—a hydrogen atom consists of one proton and one electron so that, if this electron is removed and the mass of the remaining atom determined, the mass of the proton will have been determined.

The mass of the neutron can be found by considering the conservation of momentum and energy, as in Example 17–2.

Example 17–2. When neutrons bombard protons, paraffin wax is used because it contains many protons; the protons acquire a maximum velocity of 3.3×10^7 m/s. This can be determined from the distance travelled by the protons in air when they are ejected from the paraffin wax. When the neutrons bombard nitrogen atoms, mass 14 a.m.u., the range measurements for the nitrogen give a maximum velocity of 4.7×10^6 m/s. Determine the mass of the neutron.

In a collision between neutrons and protons maximum velocity of the protons will occur when the protons are knocked straight ahead. The energies of the neutrons are so high that the chemical binding energy of the protons in the wax is insignificant. Thus applying the conservation of momentum to the collision

$$mu = mv + MV \qquad (17–3)$$

where m is the mass of the neutron, u the velocity of the neutron before the collision, v the velocity of the neutron after the collision, M the mass of the proton, and V the velocity of the proton after the collision. Applying the conservation of energy

$$\tfrac{1}{2}mu^2 = \tfrac{1}{2}mv^2 + \tfrac{1}{2}MV^2 \qquad (17–4)$$

Eliminating v between Eqns (17–3) and (17–4).

$$mu^2 = \frac{(mu - MV)^2}{m} + MV^2$$

$$m^2u^2 = m^2u^2 - 2mMuV + M^2V^2 + mMV^2$$

$$2mu = MV + mV$$

Hence, as the mass of a proton is 1 a.m.u.,

$$2mu = 3.3 \times 10^7 (1 + m)$$

For the neutrons bombarding nitrogen

$$2mu = 4.7 \times 10^6 (14 + m)$$

Hence eliminating u gives

$$m = 1.2 \text{ a.m.u.}$$

These results were the original ones used to obtain the first estimate of the mass of the neutron; more reliable measurements give the mass of the neutron as 1.00898 a.m.u. and that of the proton as 1.00759 a.m.u.

Thus an atom consists of electrons, protons, and neutrons. The greater part of the mass of the atom is

---

Experiment 17–6
Place a thin metal foil in the path of alpha particles in a continuous cloud chamber. See Experiment 16–8 for details concerning the setting up of the chamber. Observe the tracks of the alpha particles. Are many deflected in passing through the foil? Are many reflected back towards the source?

carried by the protons and neutrons. For example, oxygen: atomic number 8, atomic mass 16,

number of electrons 8
number of protons   8
number of neutrons 8

For example, iron: atomic number 26, atomic mass 56,

number of electrons 26
number of protons   26
number of neutrons 30

The great difference between the electron and the other particles is the ease with which the electron can be removed. This would seem to indicate that the electrons are less tightly held in an atom than the protons or neutrons. When we heat a filament, it is always the electrons that are released not protons or neutrons.

17–3 Nucleus

When atoms are bombarded by particles such as alpha particles, strange things happen. A typical arrangement for studying this is shown in Fig. 17–4. Alpha particles from a radioactive source hit a thin metal foil and the reflected and transmitted particles are detected, the angles at which they leave the foil being measured. Very few of the incident particles are found to be deflected more than $\frac{1}{2}°$ in passing through the metal foil. The foil would seem to be mainly empty space. (Experiment 17–6.)

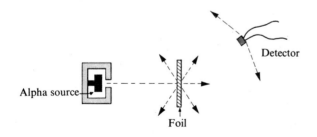

Fig. 17–4 The scattering of alpha particles

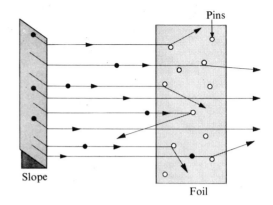

Fig. 17–5 Simple analogue of alpha-particle scattering

Only a very few alpha particles are reflected back towards the source. This, coupled with the small deviations experienced by the majority of the alpha particles, would suggest that an atom is largely empty space with virtually all the mass concentrated in a very small central **nucleus.** The alpha particles incident on the foil can be considered to be similar to ball bearings incident on rows of pins (Fig. 17–5). In order to give results comparable with those obtained in practice pins 1 mm in diameter would have to be 10 m apart. The nucleus of an atom is about 10^{-14} m in diameter, while the atom is about 10^{-10} m in diameter. These are the effective diameters determined by collisions between particles and atoms.

A possible model for the atom is therefore a small central nucleus containing the protons and neutrons with the rest of the atom consisting of surrounding electrons.

Radioactivity is a nuclear process with radiations called **alpha, beta,** and **gamma rays** emanating from the nucleus. The fact that radioactivity is a nuclear and not an atomic process can be ascertained by considering the radiation from an element when it is in various chemical combinations or physical states, e.g. different compounds of uranium in liquid, solid, and gaseous states. No effect on the radiations occurs owing to the element being in different states or compounds. As changes from one compound or state to another compound or state involve the atomic electrons and not the nucleus the obvious conclusion is that radioactive effects are nuclear in origin. How then can we obtain from a nucleus alpha, beta, and gamma radiations? Alpha-particle emission appears quite feasible when we consider that the alpha particle is just a bundle of two protons and two neutrons (a helium nucleus). Beta-particle emission, however, seems strange—there are no electrons in the nucleus. When beta emission occurs the number of protons in the nucleus increases by one, while the total mass remains constant. This can only mean that a neutron has changed into a proton and an electron, the latter being emitted as the beta particle. However, the conservation of momentum and energy does not appear to hold for such a change. The conservation laws can only be valid if another particle is also emitted, so that the energy can be shared between it and the beta particle. This other particle is called the **neutrino.** Gamma radiation is an electromagnetic wave and its emission only requires the nucleus to be in the excited state, that is too much energy. The emission is comparable with the emission of light by an excited atom. The light from an atom appears as a line spectrum due to the energy level structure of the atom; gamma radiation appears as a line spectrum due to the nuclear energy levels.

Summary

The regular arrangement of atoms in a lattice gives the structure known as a **crystal.** Such a structure can be investigated by examining the shapes of crystals by eye and also by X-rays. Atoms are made up of **electrons, neutrons,** and **protons.** The electrons carry a negative charge and have a very small mass by comparison with the proton, which has an equal positive charge, and the neutron, which has no charge but nearly the same mass as the proton. The neutrons and protons exist in the **nucleus** which occupies but a very small fraction of the volume of the atom. The effective size of atoms and nuclei can be obtained by bombardment with atomic particles.

The **atomic number** of an element is the number of positive charges carried by the atom. If the atom is neutral, it will have an equal number of protons and electrons. The **atomic mass** of the atom is the total number of the protons and neutrons when reckoned in atomic mass units; the proton and neutron each have a mass of 1 a.m.u. The arrangement of the elements in order of their atomic number is known as the **periodic table**—it is also an arrangement of the elements according to their chemical properties.

Problems

17–1 Explain how X-rays can be used to determine the structure of matter.

17–2 What are isotopes?

17–3 What evidence is there for the existence of the nucleus in an atom?

*17–4 In a 'time-of-flight' mass spectrometer all the ions are accelerated through the same potential difference. They then drift down a field-free tube and their times of arrival at the end are measured. Derive an equation relating these times to the nuclear masses.

17–5 Consider the nucleus of the atom to be represented by a table tennis ball of diameter 2·5 cm. About how far is it on this scale to the 'edge' of the hydrogen atom?

*17–6 What is the basic structure of metals?

Apparatus

The following are details of the apparatus required for the experiments. Where the item has a Nuffield Foundation Physics Project number it is given alongside the item. These items will be found in the catalogues of the major supply houses under the same number.

Chapter 1

1–1	Ticker-tape vibrator	108/1
	Tape, gummed	108/4
	12-V d.c. supply	

1–2	Ticker-tape vibrator	108/1
	Tape, gummed	108/4
	Trolley	106/1
	Runway	107
	12-V d.c. supply	

1–3	Camera	
	Xenon flasher	134/2
	or stroboscope (hand driven)	105/1
	or stroboscope (motor driven)	134/1
	Polystyrene sphere	
	Metre rule	

1–4	Spring	
	Trolley	106/1
	Foam rubber	
	Weight, 1 kg.	

1–5	Linear air track	
	Blower	
	Camera	
	Xenon flasher	134/2
	or stroboscope (hand driven)	105/1
	or stroboscope (motor driven)	134/1

1–6	Pucks kit	95
	Carbon dioxide	
	Camera	
	Xenon flasher	134/2
	or stroboscope (hand driven)	105/1
	or stroboscope (motor driven)	134/1

1–7	Trolley	106/1
	Elastic cord	106/2
	Ticker take vibrator	108/1
	Tape, gummed	108/4
	Runway	107

| 1–8 | Large trolleys | 160/1 |
| | Rope | |

1–9	Trolleys	106/1
	Runway	107
	Ticker tape vibrator	108/1
	Tape gummed	108/4

1–10	Linear air track	
	Camera	
	Xenon flasher	134/2
	or stroboscope (hand driven)	105/1
	or stroboscope (motor driven)	134/1

1–11	Linear air track	
	Camera	
	Xenon flasher	134/2
	or stroboscope (hand driven)	105/1
	or stroboscope (motor driven)	134/1
	Pin	
	Plasticine	

| 1–12 | Balloon | |
| | Rocket, water | 167 |

| 1–13 | Centripetal force kit | 172 |

| 1–14 | Centripetal force kit | 172 |
| | Spring balance | |

Chapter 2

| 2–1 | Slinky spring | 101 |

| 2–2 | Slinky spring | 101 |
| | Cotton thread | |

2–3	Ripple tank	
	d.c. supply 4V	
	12V supply for lamp	
	Cylindrical rod	
	stroboscope	105/1

| 2–4 | As 2–3 | |
| | Piece of glass or Perspex (rectangular) | |

| 2–5 | Slinky spring | 101 |

2–6, 2–7, 2–8 as 2–3

Chapter 3

3–1	Spring	
	Weights	
	String	
	Trolley	106/1
	Elastic bands	106/2
	Clock	

3–2	Sheet of paper	
	String	
	Sand	
	or felt pen and large weight	

3–3	Camera	
	Xenon flasher	134/2
	or stroboscope (hand driven)	105/1
	or stroboscope (motor driven)	134/1
	String	
	Mass	
	Trolley	
	Elastic bands	

3–4	Trolley	106/1
	Elastic bands	106/2
	Clock	

3–5	String	
	Mass	
	Clock	

3–6	Metre rule	
	Clock	
	G clamp	

| 3–7 | Capacitor, microfarads (about $1-20\,\mu$F) | |

Inductor, large, 20,000 turn
   coil on a C core
2—0—2 mA meter
Connecting wires
d.c. supply, about 4 V

3–8   Trolley   106/1
     Elastic bands   106/2
     Piece of wood, about 50 cm long
       and 10 cm wide
     Cylinder, tin tube about 4 cm
       in diameter
     Clock
     Rule

3–9   String
     Retort stands
     Weights
     Rule

3–10 Rope

3–11 Capacitor, microfarads
     Inductor (see 3–7)
     Signal generator
     a.c. meter, milliamperes

3–12 Vibration generator
     String
     Pulley
     250 g mass

3–13 Loudspeaker
     Glass tube, about 40 cm long and
       closed at one end (stand on
       bench)

3–14 Thin metal plate
     Sawdust, fine

3–15 Circular disk
     Vibration generator
     G clamp
     Headphones
     Signal generator

Chapter 4

4–1   Slinky spring   101

4–2   Slinky spring   101

4–3   Loudspeaker
     Signal generator
     Small glass beads
     Musical instruments

4–4   Loudspeaker
     Signal generator

4–5   Microphone, if not high impedance a matching transformer
     is necessary
     Cathode-ray oscilloscope
     Loudspeaker
     Signal generator

4–6   Microphone (see 4–5)
     Cathode-ray oscilloscope
     Metal reflector
     Two strips of wood

4–7   Battery-operated bell
     Bell-jar connected to a backing
       pump

4–8   Tuning fork
     Metal rod
     Wooden rod

4–9   Large trough or sink
     Tuning fork

4–10 Metal pipe, about 5 m long
     Two microphones
     Cathode-ray oscilloscope
     Two pieces of wood

4–11 Signal generator
     Loudspeaker
     Cardboard tubes
     Microphone
     Cathode-ray oscilloscope or valve
       voltmeter
     Protractor, as large as possible

4–12 Loudspeaker
     Signal generator
     Microphone
     Cathode-ray oscilloscope or valve
       voltmeter
     Balloon
     Solid carbon dioxide

4–13 Microphone

Cathode-ray oscilloscope or valve
   voltmeter
Loudspeaker
Signal generator
Two large metal screens

4–14 Two loudspeakers
     Signal generator
     Metre rule

4–15 Loudspeakers
     Signal generators
     Cathode-ray oscilloscope

Chapter 5

5–1   Two balloons
     Nylon thread

5–2   Dry battery
     Megohm resistance
     Spot galvanometer
     van der Graaf generator

5–3   van der Graaf generator
     Two metal probes, stiff copper wire
     Shallow glass dish
     Thin oil
     Fine grass seed
     e.h.t. supply (5 kV)

5–4   Bar magnet
     Iron filings
     Paper

5–5   Low-voltage power unit   104
     Copper wire
     Iron filings
     Card

5–6   Low-voltage power unit   104
     Stiff uncovered copper wire
     Magnet

5–7   Spot galvanometer
     Wire
     Magnet

Chapter 6

6–1   Mass
     Small piece of lead piping

Hammer
Battery
Wire
Matches

6–2   Cardboard cups   164
Thermometer, −10 to 110 °C
Immersion heater   75
12-V supply
Voltmeter, 12 V
Ammeter, A
    or joulemeter
Mechanical equivalent of heat
    apparatus

6–3   Aluminium block   77
Immersion heater   75
12-V supply
Voltmeter, 12 V
Ammeter, A
    or joulemeter
Thermometer, −10 to 110 °C

6–4   Continuous-flow calorimeter
Constant water head
Two thermometers
Clock
Measuring cylinder
Voltmeter, 12 V
Ammeter, A
    or joulemeter

6–5   As **6–4**

6–6   Heat of combustion apparatus
    (Nuffield Chemistry)
    or wick
Small bottle
Metal beaker or tin can
Alcohol
Measuring cylinder
Clock
Balance
Thermometer, −10 to 110 °C

6–7   Naphthalene
Test-tube
Thermometer, −10 to 110 °C
Clock
Beaker
Tripod, gauze, Bunsen burner

6–8   Two funnels
Ice
Immersion heater   75
12-V supply
Voltmeter, 12 V
Ammeter, A
    or joulemeter
Two beakers
Balance

6–9   Marbles
Tray   12

6–10   Beaker
Immersion heater   75
12-V supply
Voltmeter, 12 V
Ammeter, A
    or joulemeter
Clock
Measuring cylinder

6–11   Ether
Barometer tube
Mercury
Syringe

6–12   Filter pump
Flask
Condenser
Large bottle
Glass tubing, one piece to be about
    80 cm long
Mercury
Thermometer, −10 to 110 °C

6–13   Flask
Thermometer, −10 to 110 °C
Sodium chloride

6–14   Test-tube
Ice
Sodium chloride
Antifreeze
Thermometer, −10 to 110 °C

6–15   Ether

6–16   Boyle's law apparatus

6–17   Flask, 250 ml
Bourdon gauge

Thermometer, −10 to 110 °C
Ice
Tripod, gauze, Bunsen burner

6–18   Syringe, 50 ml
Small piece of plastic tubing with a
    small clamp
Thermometer, −10 to 110 °C
Beaker
Tripod, gauze, Bunsen burner
    or a piece of glass capillary tubing
    can be used

6–19   As **6–18**

Chapter 7

7–1   Thermistor
Avometer
    or ammeter and voltmeter with
    d.c. supply
Ice
Beakers
Tripod, gauze, Bunsen burner

7–2   Flask
Bourdon gauge
Ice
Beakers
Tripod, gauze, Bunsen burner

7–3   As **7–2**
Solid carbon dioxide
Methylated spirits

7–4   Copper wire
Constantan or iron wire
Spot galvanometer
Mercury-in-glass thermometer,
    −10 to 110 °C

7–5   Nichrome or copper wire
Thermometer, −10 to 110 °C
Metre bridge
2—0—2 mA galvanometer, jockey,
    fixed resistor, 2-V supply

7–6   12-V lamp
Rheostat, 20 Ω
Milliammeter

12-V d.c. supply
Small metal cup or crucible
Bunsen burner, tripod
Sodium chloride, potassium
  chloride, potassium sulphate

7–7 Thermometer, −10 to 110 °C
Beaker
Bunsen burner, tripod, gauze

Chapter 8

8–1 Tungsten iodide lamp   21
Screen   102
Two metal strips

8–2 Lamp, 12 V, 24 W, and power supply
Slit and holder
These and most of the other
  apparatus necessary for the optics
  experiments is available as a kit
  No. 94
Plane mirror, protractor

8–3 As **8–2**

8–4 Glass rod
Small piece of plane mirror
Light source and tube to give a
  narrow beam
Power for the lamp
G clamp

8–5 Curved mirror (cylindrical)   118
Lamp house and appropriate power
Metal comb (see **8–2**)

8–6 Line-filament bulb and appropriate
  power
Spherical concave mirror
Postcard

8–7 As **8–5**

8–8 As **8–5**
Convex mirror

8–9 Rectangular block of glass
Lamp house and appropriate power
  supply
Slit (see **8–2**)

8–10 Semicircular block of glass
Otherwise as **8–9**
Semicircular plastic cheese container

8–11 As **8–10**
Microscope slide

8–12 As **8–11**

8–13 Prism 60°
Lamp house and appropriate power
  supply
Slit
Prism 45°

8–14 Cylindrical lenses
Lamp house and appropriate power
  supply
Comb (see kit 94)

8–15 As **8–14**

8–16 Convex and concave lenses

8–17 Convex lenses of different focal
  lengths
or kit 115

8–18 Lamp house and appropriate power
  supply
Metal comb
Convex lens, about 10 cm focal
  length
Prism 60°
Postcard

8–19 Lamp house and appropriate power
  supply
Convex lens
Screen
Blue and red filters

8–20 Marble
Rigid surface, e.g. a clamped glass
  block
Protractor
Ripple tank
Curved surfaces, a curved postcard
Particle model of refraction kit 96
  or launching ramp, large ball
  bearing, two pieces of hardboard
Rectangular glass or Perspex sheet

8–21 Two launching ramps
Marbles or ball bearings
Ripple tank
Double-slits kit   97
Tungsten iodide lamp   21
Translucent screen   46/1

Chapter 9

9–1 Microphone, a matching transformer should be used if it is not
  high impedance
Cathode-ray oscilloscope

9–2 Loudspeaker
Signal generator
Ammeter
Voltmeter, 12 V
Log–log graph paper

9–3 Two loudspeakers
Two signal generators
Two ammeters and voltmeters, 12 V

9–4 Ripple tank
Dropping tube
Power supply for the tank lamp
Metal strip

9–5 Microphone (see **9–1**)
Cathode-ray oscilloscope
Two pieces of wood

9–6 As **9–5**

9–7 Motor, e.g. backing pump
Sheet of thick foam rubber

9–8 Microammeter or spot galvanometer
Strong permanent magnet
Wire

9–9 Beaker
Carbon granules
Two metal plates, about 2 cm
  square
Avometer or 2-V supply, milliammeter
Wire

12–10 As **12–9** but with a fluorescent screen instead of phototransistor, a detergent packet will do

12–11 Small X-ray tube, obtainable from Griffin & George Ltd.
Electroscope
Plastic rod
X-ray film packet
Lead foil

12–12 Ripple tank
X-ray analogue unit
Microwave transmitter
Microwave receiver
'Crystal' with associated turntable and lenses, obtainable from Unilab

Chapter 13

13–1 Conductivity kit   56
Bunsen burner
Metal cans
Large corks
Aluminium foil, Fibreglass, cotton wool
Thermometer, −10 to 110 °C

13–2 Copper bar (see experimental details)
Cardboard cups   164
Fibreglass
Immersion heater   75
Ammeter
Voltmeter, 12 V
d.c. supply, 12 V
Thermometer, −10 to 110 °C
Other bars

13–3 As **13–2**

13–4 Beaker
Bunsen burner, tripod, gauze
Potassium permanganate
Chimney model with candle
Cardboard to smoulder for smoke

13–5 Radiation kit   58
or heating element

Asbestos sheet
Glass sheet
Copper sheet
Thermometer, −10 to 110 °C
or thermopile and galvanometer
Bunsen burner

13–6 Radiation kit   58
or heating element
Variac   78
Thermometer or thermopile or phototransistor and associated apparatus

13–7 Thermocouple, i.e. copper and constantan wire
Spot galvanometer
Calibration can be in a beaker of water against a mercury-in-glass thermometer

Chapter 14

14–1 Smoke cell   93
Microscope

14–2 Kinetic theory model   11

14–3, 14–4 As **14–2**

14–5 Ball bearings
Lever balance

14–6 Plastic container   10E
Filter pump or bicycle pump
Balance
Small rectangular container   10D
Water trough
Rule

14–7 Kinetic theory apparatus   11

14–8 Balloon
Manometer
Thermocouple and spot galvanometer
Mercury-in-glass thermometer and beaker of warm water for calibration
Glass T piece

14–9 Model railway track   10R
Two carriages   10S

Two magnets
Sellotape

14–10 Bromine diffusion kit   8
Dilute ammonia
Rubber gloves
Clock

14–11 Marbles
Tray   12

14–12 Isometric graph paper   175

14–13 Bernoulli tubes   143

14–14 Kinetic theory model   11

14–15 Two thermometers
Muslin

14–16 Polythene bag, large
Two thermometers
Muslin
Calcium chloride

14–17 Test-tube
Ether
Filter pump
Glass and rubber tubing
Thermometer

Chapter 15

15–1 Demonstration diode   135
Spot galvanometer
h.t. supply and 6-V heater supply

15–2 As **15–1** or
Diode and holder   156
l.t. supply
Milliammeter
Voltmeter, 15 V
Cathode-ray oscilloscope
a.c. supply, a few volts
6-V heater supply

15–3 Deflection tube   138
e.h.t. supply
Bar magnet

15–4 As **15–3**
Ammeter
d.c. supply, about 4 V
Voltmeter if not on e.h.t. unit

15–5 Millikan apparatus
h.t. power supply

15–6 Graphite-coated paper
Metal foil
d.c. supply, 12 V
Voltmeter, 12 V
Stiff piece of copper

15–7 Cathode-ray oscilloscope
l.t. power supply, d.c. and a.c.

15–8 Demonstration photoelectric cell
h.t. supply
Spot galvanometer

15–9 Photocell 90 AV
Spot galvanometer
l.t. supply, 4 V d.c.
White light source

15–10 As **15–9**
Also filters
Rheostat
Voltmeter, 2 V

15–11 Controlled excitation tube, obtainable from Griffin & George Ltd.
h.t. supply
l.t. supply
Spot galvanometer
4Ω variable resistor
Voltmeter, 0 to 20 V
Direct-vision spectroscope or other means of observing spectra

15–12 Thyratron
h.t. supply
Voltmeter, 25 V
Milliammeter

Chapter 16

16–1 Photographic plate
Uranium compound
Developer, fixer, and facilities for processing plates

16–2 Pulse electroscope
e.h.t. supply
or gold leaf electroscope
Clock

16–3 Ionization chamber with dosimeter
h.t. supply
Clock
Uranium compound
Aluminium foil

16–4 Geiger tube
400-V supply
Cathode-ray oscilloscope
Capacitor, 500 V working
High resistance, megohms
This experiment can be performed with a Geiger tube connected to a scaler and connecting into the tube holder as in the diagram

16–5 Geiger tube and probe unit
Scaler
Clock

16–6 Radium source
Geiger unit and probe unit
Scaler or ratemeter
Paper, aluminium sheet, lead sheet
Clock

16–7 Beta source
Geiger tube
Scaler or ratemeter
Aluminium foil

16–8 Continuous cloud chamber 28
Solid carbon dioxide
Alcohol

16–9 Beta source
Geiger tube or ratemeter
Lead plates
Magnet
Clock
Alpha source
Solid-state detector

16–10 Geiger tube
Scaler
Clock
Radioactive source

16–11 Thoron generator
Cloud chamber 28
Solid carbon dioxide
Alcohol
Clock

16–12 Dice, about 100

16–13 Beta or gamma source
Geiger tube
Scaler or ratemeter
Clock
Beaker

16–14 Uranium oxide
Detergents
Cloth
Grease
Geiger tube
Scaler or ratemeter
Clock
Tray, the experiment should be conducted in a tray to avoid contamination of the bench

Chapter 17

17–1 Crystals kit 3
or various crystals such as sodium chloride, sodium thiosulphate, sugar
Microscope
Slides
Metals, galvanized steel

17–2 Calcite 3
Single-edged razor blades

17–3 Polystyrene spheres with a tray to hold them 3

17–4 Detergent
Fine tube
Beaker

17–5 As **12–12**

17–6 Continuous cloud chamber 28
Alcohol
Solid carbon dioxide

Manufacturers

The main suppliers of Nuffield type apparatus are:

Griffin & George Ltd.
Ealing Rd, Alperton, Wembley
Middlesex.

Philip Harris Ltd.
Ludgate Hill
Birmingham 3.

M.L.I. Ltd.
96-98 Putney High St,
London S.W. 15.

W. B. Nicolson Ltd.
Thornliebank Industrial Estate
Glasgow.

Certain items can be obtained from smaller concerns:

microwave apparatus and the electrical oscillations kit from Unilab, Rainbow Radio Ltd., Mincing Lane, Blackburn, Lancs.

electronic apparatus such as oscilloscopes and signal generators from Advance Components Ltd. Roebuck Road, Hainault, Ilford, Essex.

Details of suppliers of apparatus will be found in the pages of the School Science Review.

Answers to Problems

Chapter 1
1–1 10 km
1–3 If uniform acceleration 1 m/s^2
1–4 490 m
1–5 150 m
1–7 10 N
1–10 3·85 cm/s
1–11 30 N
1–15 25 N
1–16 5 J
1–17 6 J

Chapter 2
2–4 10 cm/s
2–5 2·1 mm
2–10 4 cm
2–11 2 mm
2–12 Maxima every 10 cm

Chapter 3
3–5 170 Hz
3–6 340 Hz

Chapter 4
4–3 50 cm
4–6 0·59 s
4–8 3,000 m
4–9 40 cm

Chapter 5
5–2 5 J
5–3 10^5 N/C
5–4 No
5–5 8 × 10^{-16} J
     4·2 × 10^7 m/s
5–7 2 × 10^{-4} N

Chapter 6
6–1 9·8 J
6–2 117·6 J
6–3 20,000 J, 67 W
6–4 2,100 W
6–5 1,764 J
6–7 1 × 10^4 J
6–8 2·60 × 10^5 J
6–10 4·83 × 10^7 J, 4·83 × 10^6 W
6–12 P = 160 cmHg
6–15 125 cm

Chapter 8
8–1 20°
8–6 10 cm behind the mirror
8–7 16·8 cm
8–8 1·2
    13·60 cm in front of the lens

Chapter 9
9–1 60 dB
9–2 10

Chapter 10
10–2 Factor of 4, depth of focus

Chapter 11
11–1 5 × 10^{-2} cm
11–5 13° 38′, 28° 7′, 44° 59′, 70° 27′
11–6 1·18 × 10^{-3} cm
11–7 About 5 × 10^{-5} cm

Chapter 12
12–5 14° 20′

Chapter 13
13–2 11·4 J/s, low
13–5 443 °K

Chapter 14
14–6 1·8 × 10^5 cm/s
14–7 155 °C
14–10 1 At, 27 °C; 0·76 At, −45 °C
14–11 3·3 × 10^{-22} m^2
14–12 0·060%
14–14 12 mm

Chapter 15
15–1 1·6 × 10^{-16} J
15–3 1·8 × 10^{-3} Wb/m^2
15–4 5·6 × 10^{-11} N/C
15–5 (a) Nothing, (b) no emission, (c) maximum energy of emitted electrons 0·7 eV
15–6 0 and 4·9 × 10^5 m/s
15–7 554
15–10 (a) Nothing, (b) 1,216 A emission, (c) 1,216 A

Chapter 16
16–2 1·8 × 10^5 disintegrations
16–3 (a) 50%, (b) 25%, (c) infinity, (d) 84,218
16–4 0·625 mC
16–5 2·7 × 10^9 years
16–6 350 s
16–10 15·9 years

Chapter 17
17–4 $t^2 = \dfrac{mL^2}{2qV}$
17–5 1·25 km

Index